MAKING SENSE OF MULTIVARIATE DATA ANALYSIS

JOHN SPICER

For information:

Sage Publications, Inc.
2455 Teller Road
Thousand Oaks, California 91320
E-mail: order@sagepub.com

Sage Publications Ltd.
1 Oliver's Yard
55 City Road
London EC1Y 1SP
United Kingdom

Sage Publications India Pvt. Ltd.
B-42, Panchsheel Enclave
Post Box 4109
New Delhi 110 017 India

Printed in the United States of America

Library of Congress Cataloging-in-Publication Data

Spicer, John.
Making sense of multivariate data analysis / John Spicer.
p. cm.
Includes bibliographical references and index.
ISBN 978-1-4129-0915-0 (cloth)—ISBN 978-1-4129-0401-8 (pbk.)
1. Social sciences—Statistical methods. 2. Multivariate analysis. I. Title.
HA29.S6547 2005
519.5′35—dc22

2005007119

07 10 9 8 7 6 5 4 3 2

Acquisitions Editor:	Lisa Cuevas Shaw
Editorial Assistant:	Margo Beth Crouppen
Production Editor:	Melanie Birdsall
Copy Editor:	Interactive Composition Corporation
Typesetter:	C&M Digitals (P) Ltd.
Proofreader:	Kathleen Allain
Indexer:	Kathy Paparchontis
Cover Designer:	Edgar Abarca

BRIEF CONTENTS

CONTENTS

PREFACE

The statistical techniques used in multivariate data analysis (MDA) enable the analyst to detect patterns buried in complex quantitative data. It is therefore not surprising that the techniques have proved to be of great interest and utility to behavioral and social scientists who continually grapple with highly complex phenomena. As a result, some knowledge of MDA techniques is increasingly required of students and practitioners in the behavioral and social sciences, if only to allow them to understand the research literatures in their disciplines. Courses and excellent texts on MDA abound, particularly for graduate students in these areas. What, then, is the justification for a book called *Making Sense of Multivariate Data Analysis?*

The justification for me lies in experiences over many years teaching MDA techniques to psychology graduate students and consulting on the use of these techniques as a dissertation supervisor and a journal reviewer. In my experience, many students are daunted by the prospect of engaging with MDA techniques, seeing them as an impenetrable thicket of technical complexity that they usually confront only because of course requirements. Having engaged with the techniques in some way, most leave with fragmented understandings and often a tendency to follow procedural rules in a semimindless fashion. Most striking of all, very few successfully integrate their grasp of MDA into their broader repertoire of research skills and knowledge. I believe that one major reason for these problems is that many students lack the conceptual frameworks that would enable them to appreciate the *unity* of MDA ideas and to incorporate them within their existing frameworks. This book is intended to help meet this deficiency.

The goal of making sense of MDA can be interpreted in a variety of ways. It can refer to understanding the statistical foundations of the techniques, to using computer packages that perform MDA, or to comprehending technical articles and books. There are a variety of excellent texts that provide guidance on achieving these goals, and many are cited throughout the present book. However, my objective is complementary to theirs and derives from other interpretations of sense making. In broad terms, MDA techniques rest on a few foundation blocks and gain their power through a small number of ingenious analytic strategies. So, one general way to make sense of MDA techniques is to appreciate their unity both in their foundations and their operations. The goal of this book is to highlight this unity and to do so mainly in a conceptual fashion. Accordingly, there are few statistical symbols and formulae, and there is no assumption of *any* statistical knowledge on the part of the reader.

The book is comparatively short in order to generate and maintain an overall sense of coherence and cohesion. The goal of successfully guiding readers on a rapid journey from simple foundations to complex heights is unlikely to be achieved for all readers at all times. Some will want me to accelerate or decelerate when I do the opposite. Others will balk at simplifications and insufficient qualifications of statements. Others again will want more references to support particular claims. However, a short book does not permit detailed arguments, expositions, and justifications. My aim has been to tell a coherent short story rather than to produce another lengthy textbook that painstakingly charts the MDA territory. Other writers have done this admirably, and my hope is that this book will help to make their work even more accessible.

Although there are simple structures to be found in MDA, it cannot be denied that the techniques themselves are complex and sophisticated. Accordingly, even gaining a broad conceptual overview requires the active engagement of the reader. Careful reading, rereading, and reflection are necessary if the ideas are to be assimilated. Two features of the book are intended to encourage this engagement. First, there are critical reflections on the strengths and limitations of MDA, especially in Chapter 3. The power of MDA techniques can lead the unwary to believe that the techniques are all-powerful, and from there it is a short step to misinterpretation and misuse. Confronting some of these pitfalls in an introductory text may seem premature, but I believe that establishing a critical stance from the outset is important in itself and should encourage the reader to engage in active learning.

Second, sparing use is made of figures. This may seem perverse given the traditional equation of a picture and a thousand words. However, I have found that at the introductory level of MDA, figures can be counterproductive. This is partly because of individual differences in cognitive processing, but also because the use of figures can lead the naive reader to mistake recognition for comprehension. For example, being able to draw a normal distribution or to explain it in words requires very different levels of understanding. The book therefore relies heavily on verbal exposition in order to encourage the reader to strive for comprehension rather than simple acceptance.

This book is intended as a freestanding, basic *introduction* to MDA with no prerequisites. Accordingly, it should be useful to a variety of readers and for a variety of purposes. In the context of advanced undergraduate or graduate courses, it could be used in conjunction with a set of readings or with a more advanced text. In either case, the present book could serve as a framework within which more detailed knowledge could be located and developed to the desired level. For practitioners who wish to evaluate the research literature that informs their practice, the book could provide an accessible and rapid introduction, which might be sufficient for their purposes or would more likely lead to further technique-specific reading. The book is less likely to be of interest to the experienced researcher, but it might still have some integrating function for those who feel their knowledge is fragmented. Although a psychologist wrote this book, pains have been taken to use examples from a variety of areas. So the intended audience is not limited to psychologists but is meant to encompass any behavioral or social scientist, pure or applied, who needs to become acquainted with MDA techniques.

KEY TERMS

In order to discuss the structure of the book, and given its lack of prerequisite knowledge, it is first necessary to comment on how some fundamental key terms are used. Throughout the book key terms are shown in bold when they are first discussed. Statistical analysis typically focuses on a set of **cases.** In the behavioral and social sciences, a case may be a person, a rat, a social group, an organization, a country, or any identifiable entity of interest to the analyst. The terms "case" and "individual" are used throughout the book only for consistency. Statistical information or **data** for a set of cases consist of numerical values for one or more of their attributes. Since these values typically vary across cases, a quantified attribute is commonly known as a **variable.** Some variables may be seen as somehow accounting for other variables.

The former are called **independent variables,** and the latter, **dependent variables.** For example, in Chapter 1 we will conduct a simple analysis to see whether women are happier than men. In this example, happiness is seen as a dependent variable in that it is proposed as being dependent on the independent variable of gender.

The numerical value of a variable for a case is produced by some sort of measurement process. Throughout the book we will distinguish between measurement processes that result in a **score** and those that result in a **category.** A score, for our purposes, is a number that has been produced using what is called an **interval scale of measurement.** Imagine that in the happiness example cases have given themselves a happiness score between 1 and 10. For this to count as an interval scale, we would have to assume that all of the ten possible scores are mutually exclusive, the ten scores can be meaningfully rank ordered, and the "distance" between any two adjacent scores is the same. As we will see, if these assumptions can be made, we will have a considerable head start in our attempts at statistical analysis.

Classifying the cases as men and women is an example of **categorical or nominal scaling.** To achieve this, all we have to assume is that the categories are mutually exclusive and comprehensive. (Strictly speaking, this is not measurement in the sense of assigning numbers to attributes. We could refer to men as the "1"category and women as the "2" category, but this would be arbitrary.) Distinguishing between scores and categories is important because, as we will see, their analysis requires different statistical techniques. There will be occasional reference to other sorts of measurement scales, but generally the interval and categorical scales will be sufficient for our purposes.

If one variable is analyzed on its own, the analysis is called **univariate.** If the relationship between two variables becomes the focus, the analysis is **bivariate.** When relationships involving three or more variables are analyzed, the analysis is **multivariate.** Actually, the word "multivariate" is more correctly used when there is more than one *dependent* variable, and some writers prefer the term "multivariable analysis" when this is not the case. However, in this book we will adopt the more common and looser usage of "multivariate analysis."

It is instructive to note that in the happiness example the implicit research question can be couched in two ways: Do men and women *differ* in happiness, or is there a *relationship* between happiness and gender? These sound different, but they are logically equivalent. The reason this is instructive is that it provides flexibility in how we frame an analysis. Conversely, it enables us to appreciate what apparently diverse analyses have in common. In Part 1 we will introduce simple analyses more in terms of concrete differences than of abstract relationships among variables. In particular, we will emphasize the distinction between **individual (case) differences** and **group differences.** In the happiness example, this is reflected in the distinction between how much individuals differ in happiness versus how much men and women differ. Grasping this distinction at an early stage will help to avoid confusion later.

The final term that requires preliminary comment is the notion of **accounting,** which recurs throughout the book. The objective of most multivariate analyses is to account for patterns of differences on one or more variables. Different research objectives give different meanings to "accounting." The research goal may be to predict differences, to give a causal explanation, or to reduce the set of differences to a smaller set, and this does not exhaust the possibilities. The word "accounting" is used as a generic term that is intended to cover all of these specific meanings and to carry fewer problematic connotations. As we will see in Chapter 3, confusion between the theoretical and statistical usages of a term like "predictor" is one source of misconceptions about the capabilities of MDA.

STRUCTURE OF THE BOOK

The book is divided into two parts: The three chapters in Part 1 introduce the core ideas of MDA, and the five chapters in Part 2 explore the techniques themselves. Chapter 1 introduces the univariate and bivariate statistical building blocks that form the foundation of MDA, using small, contrived data sets. As noted earlier, the discussion assumes no prior statistical knowledge. In Chapter 2 we review the variety of factors that can influence the trustworthiness of results from any statistical analysis, in particular the role of chance. Then, in Chapter 3 we examine the reasons why MDA is needed and the strategy that lies at the heart of any MDA technique. This is the notion of the composite variable, whereby multiple variables can be packaged into one. Finally, in Chapter 3 we stand back from the statistical details and reflect on the strengths and limitations of MDA techniques. Together, these three chapters provide the foundation on which Part 2 is built and should therefore ideally be read first and in order. Even the reader with a good grasp of basic statistics should find material worth pondering on.

The five chapters in Part 2 introduce six MDA techniques: multiple regression, logistic regression, discriminant analysis, multivariate analysis of variance (MANOVA), factor analysis, and log-linear analysis. There are a number of issues dealt with in detail in Chapter 4 on multiple regression to which frequent and briefer reference is made in later chapters. Accordingly, while the chapters in Part 2 could be read in any order, it is probably easier to read Chapter 4 first. Most of the technique chapters share a broad common structure. This begins with a discussion of how the composite variable strategy is used within a particular technique. Then the particular statistical tools for the technique are introduced and their use is illustrated with actual data and examples from the research literature. After this, the issues affecting the trustworthiness of results produced by the particular technique are reviewed. The main exception to this structure is in Chapter 6, where it is necessary to provide a fairly lengthy preamble on the analysis of variance before introducing its multivariate counterpart MANOVA.

The techniques covered in Part 2 are a selection of those found in the MDA area. Since one of the primary aims of the book is to demonstrate the continuity between basic statistical building blocks and multivariate analysis, more advanced techniques such as path analysis and structural equation modeling have been excluded. These techniques sit on the next story up in the building that is statistical analysis and are better left for more advanced and comprehensive texts. Even on the first level of MDA techniques, there are too many to be included in a short introduction. Those selected are a variety in a number of senses. Four of the six techniques are used to examine how multiple independent variables account for one or more dependent variables. But in the last two chapters, the independent/dependent variable distinction disappears, and other analytic objectives emerge. The techniques also vary in the ways in which they deal with scores and categories, as signaled earlier. Finally, the techniques vary in their frequency of usage in the behavioral and social sciences. Multiple regression and factor analysis are the workhorses of countless nonexperimental studies, while MANOVA is frequently used in experimental studies. In contrast, logistic regression is rapidly increasing in popularity and overtaking discriminant analysis, which is used for the same analytic purposes but is more problematic. Finally, log-linear analysis, which focuses on categorical data, is probably the least used but deserves wider appreciation and adoption.

All of the examples in the book are concerned with what psychologists call "subjective well-being" or, more loosely, happiness. This choice was partly motivated by the desire to accentuate the positive in a context that is often associated with negative feelings. But the focus was also chosen because this topic is an intrinsically interesting and burgeoning one in many parts of the behavioral and social sciences. One piece of research to which we return repeatedly in Part 2 is a study of the effects of workplace characteristics on the well-being of nurses. This is a published study (Budge, Carryer & Wood, 2003), but the authors have also kindly provided me with their data for detailed analyses using all of the MDA techniques covered in Part 2. All of these analyses were conducted with the SPSS for Windows package, Version 11.5, which is one of the most popular statistical packages among behavioral and social scientists.

At the end of each chapter are suggestions for further reading. To facilitate the flow of the text, references within chapters have been kept to a minimum. Almost all of the references deliberately guide the reader to sources that are accessible for the beginner. Those sources in turn provide guidance toward the more technical literature for those who wish it.

ACKNOWLEDGMENTS

I am very grateful for the contributions made by a number of people in the preparation of this book. The suggestions made by six anonymous reviewers at different stages were much appreciated. Inevitably, I did not always follow their advice, and any remaining errors or patches of fog are my responsibility. I am also very grateful to the literally thousands of students who over the years have pushed me to find new ways of understanding and communicating complex ideas. In particular, I thank those who encouraged me to write a book on multivariate data analysis. Turning to those contributors who can be named, I have been very fortunate to benefit from the outstanding editorial support provided by Lisa Cuevas Shaw. Her enthusiasm and professional acumen throughout the process have been invaluable. A debt of gratitude is also due to Claire Budge, Jenny Carryer, and Sue Wood, who very generously allowed me to dissect their published work and data as a running example in the book. Finally, more than tradition leads me to reserve my deepest appreciation for my wife, Claire. As a research psychologist and methods enthusiast, she has acted as a critical sounding board during countless hours of reading and discussion. As a partner, she has continually provided the many types of support that authors need to succeed in their solitary projects. This book is because of Claire, and for Claire.

PART I

THE CORE IDEAS

ONE

WHAT MAKES A DIFFERENCE?

At first sight multivariate data analysis (MDA) can appear diverse and impenetrable. However, it has two striking and reassuring features that will enable us to make sense of MDA techniques more easily than might be expected. Multivariate methods can be understood as logical extensions of simpler techniques, and further, they rely on the same small set of "moves" to achieve their ends. In the first two chapters we will explore the relatively simple statistics that later become the building blocks of multivariate techniques. As noted in the preface, the discussion in these chapters will assume no prior statistical knowledge and will attempt to provide an intuitive rather than a technical grasp. (For readers who have very little or no prior statistical knowledge, it would be helpful to review the section on key terms in the preface before proceeding.) In Chapter 3 we will examine the strategies that lie at the heart of multivariate techniques. Again, the aim is to make sense of these strategies in terms of their logic rather than the statistical technicalities that lie beneath.

In the following sections we approach the chapter title in two different ways. First we ask how differences can be quantified or, more precisely, how a single set of differences can be summarized numerically. Then we examine how the relationship between two sets of differences may be analyzed. To the extent that a systematic relationship is thereby detected we then have some evidence that one set of differences may account for the other. We will explore these two issues of quantifying and accounting for differences twice over: first in Section 1.1 for data in the form of scores and then much more briefly in

Section 1.2 for data in the form of categories. Since the main aim in the first two chapters is to expose the building blocks of MDA and to ground them firmly, we will make extensive use of small contrived examples. In Part 2, where we explore multivariate techniques as such, we will turn to examples of real-world data drawn from a variety of sources.

1.1 ANALYZING DATA IN THE FORM OF SCORES

1.1.1 Univariate Analysis: Capturing Differences in One Set of Scores

Imagine that we ask five individuals, conveniently named A, B, C, D, and E, to complete a measure of subjective well-being or happiness in which the possible score range is between 1 and 5 and a higher score indicates a higher level of well-being. Even more conveniently, imagine that each of these five individuals scores differently on the well-being measure. This situation is shown in Figure 1.1, where the individuals occupy the rungs of a score "ladder." Person A scored 5, person B scored 4, and so on. How can this set of different scores be summarized numerically to indicate just how different these individuals are from each other? In addressing this question, we are particularly interested in finding so-called summary statistics that are elegantly simple and that have the potential to be used as building blocks in more complex situations.

Even with only five cases, wondering about how to summarize all of their pairwise differences is hardly a simple beginning and in fact leads nowhere very helpful. Instead we begin by focusing on how much each person's score differs from a fixed reference point. For the purpose of developing building blocks, it turns out that the **arithmetic mean** is a particularly helpful reference point. The arithmetic mean is just the sum of all scores divided by the number of scores. Here the mean is 15/5 = 3, a number that identifies the midpoint of the scores in a way we will discuss later. So now we can re-express each person's score as the amount by which they "deviate" from the mean of 3. These **deviation scores** are shown to the right of the ladder and indicate that person A and person E are the most deviant, scoring 2 above and below the mean, respectively. Person C is totally nondeviant, but notice that often in practice there are no individual scores at the mean, which in fact may not be a possible score at all if it is not a whole number. As we make use of summary

Well-Being Score		*Deviation Score*
5	A	+2
4	B	+1
3	C	0
2	D	–1
1	E	–2

Figure 1.1 Well-Being Scores for 5 Individuals (A–E)

statistics such as the mean we leave the cases behind and shift to a different aggregate level of description. This issue of multiple levels of analysis is an important one that we will consider further at the end of Chapter 3.

So far we have simply replaced the five original scores with five deviation scores and so have yet to *summarize* differences in any way. Calculating the mean suggested that adding up scores is a useful summarizing strategy; five scores were replaced by one statistic that captured the middle of these data in one sense. Just in what sense becomes clear when we try the adding strategy with deviation scores and find that they will always sum to zero. This is because the mean has the interesting property that the deviations above and below it will always cancel out exactly. It is the fulcrum around which the scores will always balance. What to do after such a promising start? As usual we choose the option that has the most potential for acquiring useful building blocks. In the present situation the trick is to sum, not the deviation scores, but the *squared* deviation scores. Whether a number is positive or negative, multiplying it by itself results in a positive number, so the problematic canceling out of positive and negative deviations disappears. For the numbers in Figure 1.1, the sum of the squared deviation scores is 4 + 1 + 0 + 1 + 4 = 10. This statistic is known as the **sum of squares** (shorthand for the sum of squared deviations around the mean) and is a true cornerstone of multivariate data analysis, as we will see.

The sum of squares captures the total amount of differences or variability in the data, but how might we summarize the *average* amount of difference? As with the mean, we can simply divide the total by the number of data points, that is, 10/5 = 2. This average sum of squares is known as the **variance,** another cornerstone of statistical data analysis. Calculating it in this straightforward way is appropriate in some circumstances, but with a slight adjustment we can arrive at a form of the variance that will serve many more purposes in more complex analyses. The adjustment involves dividing the sum of squares, not by the number of data points (5 in this case), but by the number of points that are free to take on any value. This adjusted divisor is known as the **degrees of freedom,** a notion we will encounter repeatedly but one we will not need to understand in detail. In the present case the degrees of freedom are 5 – 1 = 4. This is because, once we have fixed the value of the mean in our calculations, the final data point is constrained to take on the value that will result in this outcome. All of the other four scores can take on any value in the range allowed by the measure, but the fifth score has no such freedom. This no doubt sounds rather mysterious, and in fact the concept of degrees of freedom really only begins to take hold when we treat cases as a sample drawn from a population—a major topic in Chapter 2. For now, the key point to note is that the variance is a fundamental measure of average difference or variability, and it always comprises a sum of squares divided by a degrees of freedom, in this case, the number of data points minus 1. The variance for the example data is therefore 10/4 = 2.5.

In the variance we have a useful number that summarizes the average amount of differences in a set of scores. One of its limitations, though, is that by squaring scores we have moved away from the original scale. The original scores are on the 1–5 well-being scale, but the variance is in squared well-being units—a surreal and not very helpful scale. To remove this dislocation, we can simply "unsquare" or take the square root of the variance to produce a statistic called the **standard deviation.** For the present data, the square root of 2.5 is approximately 1.58. So if we want to summarize the scores in Figure 1.1 succinctly, we can simply note that there are five scores with a mean of 3 and a standard deviation of 1.58. A particular variance or standard deviation conveys useful information, but they really come into their own when used as building blocks. Before we start to build, it will be helpful to get a feeling for the features of data that influence the values of the mean, variance, and standard deviation. To do this, we turn to the three sets of data shown in Figure 1.2.

We have now extended our imaginary alphabetic sample to 10 individuals (A–J) and obtained their scores on three different measures, each of which has a possible score range of 1–5. The left-hand ladder contains their scores on a well-being measure, the middle ladder shows their scores on a measure of positive emotions (referred to as positive affect), and the right-hand ladder

	Well-Being	Positive Affect	Life Satisfaction
5	A	A	AB
4	BC	B	CDE
3	DEFG	CDE	
2	HI	FGHIJ	FGH
1	J		IJ
Mean	3.00	2.80	3.00
Variance	1.33	1.07	2.44
SD	1.16	1.03	1.56
Median	3.00	2.50	3.00

Figure 1.2 Three Sets of Scores for 10 Individuals (A–J)

contains their scores on a measure of satisfaction with life. The data are again depicted by placing individuals on the ladder rung corresponding to their score. The mean, variance, and standard deviation (SD), and another statistic called the median, appear beneath each ladder.

The well-being scores on the left-hand ladder have a mean of 3, a variance of 1.33, and a standard deviation of 1.16. Now that we have more individuals we can start to reflect on how they are distributed on the ladder, that is, the frequency distribution of their scores. The shape of the left-hand configuration is referred to as a **normal distribution,** meaning that most individuals appear in the middle of the distribution with decreasing frequencies for the higher and lower values. A further key feature of normality is that the distribution is symmetrical about its center, that is, the lower rungs are a mirror image of the upper rungs. The closer a distribution is to normal, the more helpful are the mean, variance, and standard deviation as indicators of the middle and spread of the distribution.

The distribution on the middle ladder of positive affect scores suggests that we have happened upon a group of people who are not very joyful. Their scores have a mean of 2.8, a variance of 1.07, and a standard deviation of 1.03, all lower than those for the left-hand distribution. So, as we can see, compared with the left-hand distribution, the center has dropped down the ladder a little, and the individuals show fewer differences as they cluster more tightly on rungs 2 and 3. (Note that comparing means, variances, and standard deviations across distributions in this way can be done only if the two sets of scores are measured in the same units since all three statistics are "scale bound.") Whatever else we might say about the middle distribution, it is clearly nonnormal. Most individuals appear at the lower end of the ladder and the distribution is obviously not symmetric about its center. This asymmetry is referred to as a **skewed distribution,** more precisely a positively skewed distribution, because the scores "tail off" at the high or positive end of the scale. Were the tail pointing in the opposite direction, the distribution would be negatively skewed. Sometimes the asymmetry is due less to a continuous tail than to a few individuals whose extreme scores locate them well away from the crowd. Such scores are called **outliers,** which can influence the variance and standard deviation in a powerful way because of the magnifying effect of squaring deviation scores.

Whether the nonnormality of a distribution is due to a skewed tail or to outliers, both have a magnetic effect on the mean, dragging it away from the center of the data and rolling on to influence the values of the variance and standard deviation. The distorting effect on the mean can be seen if we compare the means for the left-hand and middle distributions with their corresponding medians. The **median** is the score that splits a sample in two, with half of the individuals above and half below. Unlike the mean, the median does not make use of the score values as such and so cannot be influenced by outliers or skewness. For the left-hand distribution, the mean and median are both 3, indicating no distortion. But in the middle distribution the mean of 2.8 has been dragged above the median of 2.5 because of the positive skew. In this constructed example the degree of skewness and its effect on the mean are not great, but in practice skewness and outliers can cause major distortions that then ripple through any analyses that have the mean at their heart.

Turning finally to the right-hand ladder of satisfaction scores, we see that the mean is 3 (as is the median), the variance is 2.44, and the standard deviation is 1.56. These figures are reassuring in that the mean clearly sits in the center of the distribution, and the variance and standard deviation show that individual differences are more pronounced than in either of the other two distributions. However, these numbers are not to be trusted because they hide a crucial aspect of the data. The distribution is symmetrical, but it is not normal because it does not have one center but two. Put more technically, the distribution has more than one **mode** or peak; it is bimodal with peaks for the 2 and 4 score values. In practical terms this may be signaling measurement problems in the midrange or the presence of two subsamples. But for now the more important statistical point is that the mean, variance, and standard deviation are not designed to work with distributions where there are multiple modes or peaks. So, in summary, these statistics are most effective when they are applied to a normal frequency distribution, that is, one that is symmetrical about a single peak. In Figure 1.2 the normality of the left-hand distribution makes it a good candidate for these summary statistics, the skew in the middle one threatens some distortion, and the bimodality in the right-hand distribution undermines the use of the statistics completely.

Before we start to build with the blocks of the mean, sum of squares, variance, and standard deviation, two concluding comments are in order—a caveat and another way of thinking about these statistics to pave the way for things to come. It is important to appreciate that these summary statistics can be calculated and interpreted in a meaningful way only if the scores are measured on at least an **interval scale.** In other words, as we noted in the preface, the scale values must be mutually exclusive, in rank order, and equally spaced. Clearly, if the distance between, say, scores of 1 and 2 were not the same as that between scores of 4 and 5, then the arithmetic operations we have conducted would unravel. In terms of the ladder image, the rungs have not only to be fixed, but also to be fixed with equal spaces between them. We will return to this and other measurement issues in Chapter 2.

We can begin to open up another perspective on these statistics by raising the following question: If we had access to summary statistics but not the individual scores for a group of people, what would be our best guess for any given individual's score? One answer to this question is to always choose the mean score for the group. To understand why, we need to rethink the distance between any individual's score and the mean as the amount by which our mean-inspired guess has failed. Viewed in this way, the sum of squares is the total amount by which the mean "misses" individual scores. If everybody scored at the mean, there would be no misses, but obviously this is exceedingly rare. A key property of the mean is that if it is used as the reference point in calculating individual differences, it will produce a smaller sum of squares than will any other reference point such as the median. So the mean is a best guess of individual scores because it minimizes the miss rate, as long as we think of misses in terms of squared distances from the mean. Details aside, the key general idea to note for now is that guesses or predictions about individual scores based on averages almost always fail to some extent. The extent of this failure can be quantified using the sum of squares, variance, and standard deviation. So these statistics can be used as indicators not only of individual differences, but also of error.

1.1.2 Bivariate Analysis: Accounting for Score Differences With Categories

We have explored only a few building blocks so far, but already we can start to put them to work in interesting ways. Returning to the 10 individuals shown in the left-hand ladder in Figure 1.2 and reproduced on the left-hand side of Figure 1.3, we can speculate about what factors might account for differences in well-being and now actually undertake an analysis to evaluate the speculation. There is some evidence that women are likely to experience higher levels of well-being than men (Wood, Rhodes & Whelan, 1989). Is this difference evident in the data shown in Figure 1.3? Now we are treating well-being as a dependent variable and gender as an independent variable. Gender is a **categorical variable** in that it simply assigns individuals to unordered categories, two in this case. We can analyze the relationship between a dependent variable consisting of scores and an independent variable consisting of categories, using simple extensions of the statistics we now have at our disposal.

Figure 1.3 shows the well-being scores for the 10 individuals split into two groups: 5 women on the middle ladder and 5 men on the right-hand ladder. Underneath the ladders are the means, sums of squares (SS), variances (Var), and standard deviations (SD) for each grouping.

The similar variances and standard deviations indicate that there is a very similar amount of individual differences or variability within each of the three groupings. Note that the amount of variability in the women and men subgroups is identical because the two distributions happen to be mirror images of each other. But the question we really want answered is whether men and women differ in well-being as *groups,* not as individuals. The answer to this clearly lies in the *difference* in their mean scores: 3.4 – 2.6 = .8. This difference suggests that at least in this sample of 10 people, women report higher levels

	All Individuals		*Women*		*Men*
5	A	5	A	5	
4	BC	4	B	4	C
3	DEFG	3	DE	3	FG
2	HI	2	H	2	I
1	J	1		1	J

	All Individuals	Women	Men
Mean	3.00	3.40	2.60
SS	12.00	5.20	5.20
Var	1.33	1.30	1.30
SD	1.16	1.14	1.14

Figure 1.3 Well-Being Scores for 5 Women and 5 Men

of well-being on average. This is a clear outcome, but how might we extend our simple statistics to dig deeper and open up analytic possibilities for more complex and realistic situations in the future?

The key point to note first is that we now have analyses going on at two levels: the level of individual differences and the level of group differences. Put another way, we are now quantifying **within-group** and **between-group differences.** The next step is to ask how we might combine these two sorts of information in helpful ways. It would be particularly helpful to be able to compare them and to ask whether the gender difference in well-being is greater than we would expect on the basis that individuals differ from each other anyway. In other words, is the between-group difference more than we would expect simply from within-group differences? We could explore this question at this point by considering techniques that focus only on a difference between two groups. But, as usual, we want to develop strategies that can subsequently be extended into more complex situations such as those involving more than two groups and multiple independent and dependent variables. Accordingly, we now introduce a technique called the **analysis of variance (ANOVA),** which appears in many forms throughout the realms of data analysis.

Earlier we found that a helpful way to capture individual differences was to begin with the notion of deviations from the mean and then develop this into the sum of squares and variance statistics. We can use exactly the same approach to capture differences between *groups.* Instead of focusing on mean differences as such, we calculate how far each mean deviates from the mean of the means, that is, the mean for the whole group. Looking back at the means in Figure 1.3, we see that the women's mean is .4 above the mean for the whole group and the men's mean is .4 below it. If we square each of these and sum the results, we arrive at a sum of squares that captures how different the means are. However, later we will want to compare these between-group differences with within-group differences. At the moment the two would not be comparable because one uses groups as the unit of analysis and the other uses individuals. The obvious way out of this is to convert the group calculations back to the individual level by multiplying each squared deviation score for a group by the number of cases in that group. So, the **between-groups sum of squares** would be calculated as $5(+.4^2) + 5(-.4^2) = 1.6$. This number translates the mean difference into a more flexible statistic that can be subsequently compared with individual level differences and applied to any number of means. As in the case of the sum of squares for scores, we can convert the between-groups sum of squares into a **between-groups variance.** We do this as before by dividing it by the appropriate degrees of freedom: the number of data points minus 1. Since the between-groups sum of squares is based on two means, the **between-groups degrees of freedom** is $2 - 1 = 1$. So the between-groups variance is $1.6/1 = 1.6$.

Soon we will pull all of these elements together into a coherent and hopefully satisfying pattern, but there is one more step before we do. We have discussed how to quantify individual differences for the group as a whole and the mean difference between the subgroups of women and men using appropriate sums of squares and variances. What about the differences *within* the subgroups of women and men? The subgroup sums of squares are shown under the respective ladders in Figure 1.3. These can simply be added together or pooled to produce the **within-groups sum of squares:** $5.2 + 5.2 = 10.4$. As usual, this can be converted into a **within-groups variance** by dividing by the appropriate degrees of freedom. Each group has $5 - 1$ degrees of freedom, and these are again pooled to produce the within-groups degrees of freedom of $4 + 4 = 8$. So the within-groups variance is $10.4/8 = 1.3$.

This completes all of the calculations we need to provide a comprehensive summary of the differences shown in Figure 1.3. The results of an analysis of

Table 1.1 ANOVA Summary Table Showing the Relationship Between Gender and Well-Being

Source of Differences	*Sum of Squares*	*Degrees of Freedom*	*Variance*
Between groups	1.60	1	1.60
Within groups	10.40	8	1.30
Total	12.00	9	

variance, however simple or complex, are conventionally shown in a summary table like that in Table 1.1. This simply arranges the statistics we have calculated in a convenient pattern and so contains no new numbers.

Before we review the numbers and make use of them, a few comments on alternative terminology in this sort of summary table will be helpful. It is common for the first column to be headed "source of variability." "Source of differences" is used here instead to keep the discussion consistent, although the terms are logically equivalent. The label "within-groups" is sometimes replaced by "error," which should at least resonate with the earlier comment on variability as an indicator of failure to predict. Finally, variances in this context are usually referred to as "mean squares." This is a more informative label since it refers to the mean sum of squares, but again the "variance" label has been kept for consistency and to minimize confusion. So encountering expressions such as "error mean square" should not be a cause for confusion but an occasion for translation into the more familiar "within-groups variance."

The three rows in the summary table highlight the fact that we have analyzed the well-being differences in three different ways. In the bottom row we find the statistics that index differences in all of the cases: the total picture. This total has been split into differences that can be accounted for by being in the men's or women's groups: the between-group differences and the within-group differences, which are just individual differences. Notice that in the case of sums of squares and degrees of freedom, this split is additive, that is, the between- and within-group numbers literally add up to the total. Sometimes this is referred to as "partitioning" or dividing up the total variability into its components. This additive property will turn out to be especially valuable when we later try to make sense of complex sets of differences. Notice also that variances are not additive, which is why there is no variance entry in the "total" row. As we saw in Figure 1.3, the total variance for the whole group is

1.33, but this is not the sum of the between- and within-group variances, and so to include it in the summary table could be misleading.

Having taken the trouble to calculate these statistics, how might we use them to shed further light on the question of whether and how men and women differ in their well-being scores? Earlier it was suggested that we might compare the between- and within-group variances to see if the former was sufficient to "rise above" the latter. This can be done by dividing the between-groups variance by the within-groups variance to produce a statistic called the ***F* ratio.** In the present case this would be 1.6/1.3 = 1.23. The more this rises above 1, the more evidence we have that between-group differences are present. However, it cannot be interpreted as a direct measure of the *amount* of group difference, which is still best captured by the actual mean difference or some derivative of it. The *F* ratio is indicative of between-group differences, but we will postpone consideration of its more legitimate uses until Chapter 2. This is also why it has been omitted from the summary table, where it is usually shown in another column on the right-hand side.

We can combine elements from the summary table to produce statistics that summarize individual differences. One of these is called **eta^2,** which is found by dividing the between-groups sum of squares by the total sum of squares: 1.6/12 = .133. When this is multiplied by 100, it can be interpreted as the percentage of total variability accounted for by the categorical variable. So, in the present case we conclude that gender accounts for 13.3% of the variability or individual differences in well-being scores in this sample of people. Alternatively, we could describe the situation in terms of *unexplained* variability by computing a statistic called **Wilks's lambda,** which is the ratio of the within-groups and total sums of squares. This is 10.4/12 = .867 and indicates that gender *fails* to explain 86.7% of the variability or individual differences in well-being scores. Clearly, eta^2 and Wilks's lambda are mirror images of each other and by definition their values always add up to 1 or 100 in percentage terms. Notice again that both of these statistics reflect individual differences. Only mean differences, or some statistic derived directly from them, convey what is happening at the group level.

This completes our introduction to some simple statistics that allow us to examine whether category differences can account for a set of score differences. By now you should already have some sense of how much analytic work can be done with a few simple building blocks. In Chapter 6 we will extend these ideas to much more complex situations where we encounter so-called **multivariate analysis of variance.** Before that, Chapter 5 will introduce another multivariate technique called **discriminant analysis,** which is essentially another approach to multivariate analysis of variance that reverses the status of independent and dependent variables. So there we will be exploring how to account for category differences with scores and appreciating that many statistics are blind to the independent and dependent status that researchers choose for their variables. Notice, for example, how in the data we have been considering the statistics allow us to say either that gender accounts for 13.3% of the variability in well-being or that well-being accounts for 13.3% of the variability in gender. Alternatively again, we could accurately say that gender and well-being *share* 13.3% of their variability. The independent and dependent status of the variables comes from the research context, not from the statistics.

1.1.3 Bivariate Analysis: Accounting for Score Differences With Scores

In the previous section we conducted a simple analysis of variance to explore how far gender might account for differences in well-being. Now we return to the data on positive affect shown in Figure 1.2 and ask to what extent differences in well-being might be accounted for by differences in positive affect. We will continue to assume that positive affect was measured on an interval scale, and so now both independent and dependent variables are in the form of scores. To analyze whether score differences on a dependent variable can be accounted for by scores on an independent variable, we turn to a technique called **simple regression.** Since analysis of variance is a special case of regression, we can introduce the latter with very few new ideas or tools.

To build an initial bridge between analysis of variance and regression, it will be helpful to reconfigure and redescribe part of Figure 1.3. Figure 1.4 again shows the distribution of well-being scores for women and men separately. The distribution for the whole group has been removed, as have all of the ladder frames. The means for the women and men are shown as circles joined by a solid line. The mean for the whole group is shown as a dotted horizontal line. So nothing has changed from Figure 1.3 other than the depiction, which is now referred to as a **bivariate scatter plot** as it summarizes the relationship between two variables. As we saw earlier, at the group level the relationship can be described as the difference between the two means. These

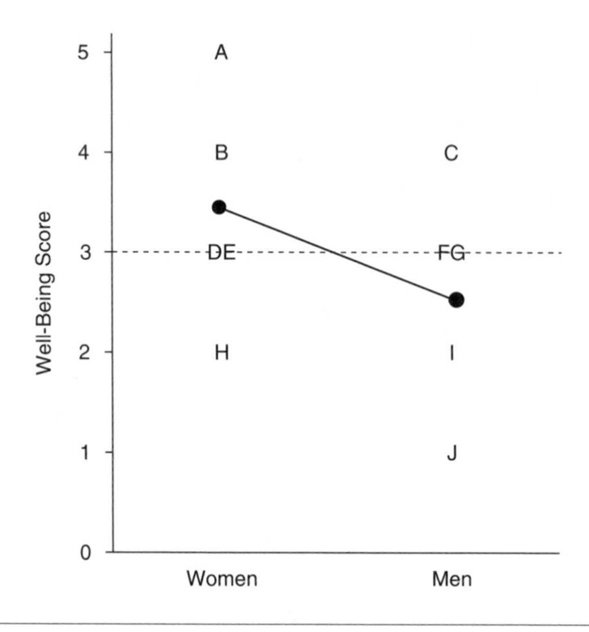

Figure 1.4 Scatter Plot Showing the Relationship Between Gender and Well-Being

group means are also referred to as **conditional means** because their value on the dependent variable of well-being is conditional on the particular category of the independent variable. Another way to describe the group difference is in terms of the **slope** of the solid line that joins the means. This drops from 3.4 to 2.6, so the slope is –.8. The slope has a negative value because it shows a *decrease* in well-being as we travel from left to right along the horizontal axis. Finally, the group difference can also be expressed in terms of the vertical distances between the conditional means and the mean for the total group shown by the dotted line. Remember that these distances can be squared and summed to enter into the between-group sum of squares.

With these perspectives in mind, we can now focus on a regression approach to the question of whether positive affect levels might account for differences in well-being. Figure 1.5 is a scatter plot that reconfigures the well-being and positive affect data that we saw in Figure 1.2.

The dependent well-being scale appears on the vertical axis, known generally as the **Y axis,** while the independent positive affect scale appears on the horizontal, or **X, axis.** Although they are not shown as such, we now have five ladders, one for each value of the positive affect scale. In Figures 1.3 and 1.4

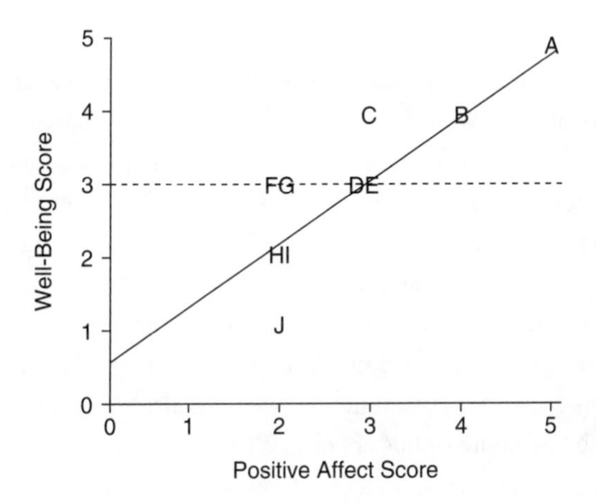

Figure 1.5 Scatter Plot Showing the Relationship Between Positive Affect and Well-Being

we split up the total distribution of well-being scores according to the categories of the independent variable—women and men in that case. The same process has occurred in Figure 1.5, where individuals appear in vertical subgroups according to their positive affect score. This means that each letter now represents an individual's *pair* of scores: person H scored 2 on both measures, person C scored 3 on positive affect and 4 on well-being, and so on. Again the dotted horizontal line shows the mean well-being score for the whole sample. Of most importance for present purposes, the solid line again captures the relationship between the two variables at the group level and is generally known as the **regression line.** In the following we will first look more closely at this line and its interpretation as the *impact* of positive affect on well-being. Then we will bring back the analysis of variance perspective and use it to explore the relationship of these variables at the level of individual differences. Finally, we will look inside the new statistics we have encountered to see how they all make use of the same few building blocks.

The Regression Line

We just noted that the solid regression line in Figure 1.5 represents the impact of positive affect on well-being in this sample of 10 individuals. To see why, it is helpful to ask first where the line comes from. A logical extension of the strategy we used in Figure 1.4 would be to draw a line that joined up the conditional means, that is, the mean well-being scores for the subgroups of individuals at each value of the positive affect scale. However, these conditional means are not well defined. There is only one individual (A) in the 5 column, one (B) in the 4 column, and none at all in the 1 column. Moreover, in our continuing search for elegance and simplicity, it would be satisfying to summarize the relationship with a straight line—the simplifying assumption of **linearity.** So instead of joining up the conditional means, we need an alternative approach that nonetheless produces a similar outcome.

The answer is to make use of an approach based on the principle of **least squares.** This may sound daunting, but we have already encountered the principle near the beginning of this chapter. When we were summarizing one variable, we chose to use the mean as a reference point and best-guess statistic. A justification for this was that the mean resulted in a sum of squared deviations smaller than that produced by any other reference point. In other words, the mean is a desirable statistic because it obeys the principle of least squares: Its numerical value depends on minimizing the sum of squares around itself. The regression line can be thought of as a mean stretched across two dimensions. A set of data points in one dimension (one ladder) can be summarized with a point—the mean. The relationship between data points in two dimensions can be summarized with a line—the regression line. The chosen line is that which minimizes the vertical distances between itself and all of the individual data points. So the regression line cuts through the middle of the data points in the sense that it follows the straight path that produces the smallest sum of squared deviations around itself. In Figure 1.5, for example, we might be tempted to draw the regression line more steeply to get closer to individuals F, G, and C. But this would increase the distances from the other 7 individuals, especially the distant J, and result in a larger sum of squared vertical distances from the line.

Statisticians have devised computations that identify the regression line for a set of data, and we will look at some of these later. This suggests that the line can be expressed as numbers, but what are they? A straight regression line can be uniquely identified with two numbers. The first we have already introduced as the **slope** of the line. This number indicates how steeply the line goes up or down by telling us how the Y (dependent variable) value changes on its scale when the X (independent variable) value changes by 1 unit on its scale. The slope for the Figure 1.5 regression line turns out to be approximately .94. So this indicates that as the positive affect score increases by 1 unit, the happiness score increases by .94 units—almost a one-to-one relationship.

The slope identifies a family of parallel lines rather than one line in particular. To pin down the unique line for the data in Figure 1.5, we need a second number: specifically the number that tells us where the line meets the upright axis. This is known as the **Y intercept** or sometimes as the constant. Its value is approximately .37: the point on the well-being scale where the regression line meets or intercepts the Y axis in Figure 1.5. The slope and the Y intercept jointly define a unique regression line and are known generally as **regression coefficients.** They can be combined into a single **regression equation** that summarizes the relationship between any X and Y variable. Generally, this relationship is expressed as predicted Y = slope(X) + Y intercept. In words this says that the Y value for any individual can be predicted by multiplying that individual's X score by the slope and adding the product to the Y intercept. For example, the predicted well-being score for individual B who scored 4 on the positive affect measure would be (.94)(4) + .37 = 4.13.

Since the slope and Y intercept define the regression line, we can translate this idea back into graphical form. Looking at Figure 1.5, we can again make a prediction for individual B by noting what well-being score corresponds with a positive affect score of 4 according to the regression line. An imaginary line extending up from a positive affect score of 4 hits the regression line at a well-being score of just above 4, as we just calculated. This predicted score is very close to person B's actual score of 4. But notice what happens when we try the same procedure for person J with a positive affect score of 2. The predicted well-being score according to the regression line is about 2 (more precisely 2.25), but the actual score is 1. The difference between the predicted and actual scores is known as the **residual**—a statistic we will have much more to say about later. This procedure highlights the important point that we are using group summary statistics—the slope and Y intercept—to make individual predictions. Just as when we treated the mean as a basis for prediction, we are only partially successful. Some of the differences in well-being scores are explained or predicted, some are not. Again, notice how we continually move around between-group and individual levels of analysis.

The slope and Y intercept carry very different interpretative weight. The slope is the key statistic that captures the impact of an independent on a dependent variable. If someone asks us how far differences in positive affect account for differences in well-being in the present group, the slope of .94 precisely answers that question *at the group level of analysis.* As we noted, it says that a 1-unit increase in positive affect is associated with almost a 1-unit increase in well-being on average. This slope has an implied positive sign, so it signifies that positive affect and well-being scores move up and down in concert. But a slope may have a negative value, which would mean that as the X values increased, the Y values decreased, that is, they would move in opposite directions. In general, note that a higher slope value, positive or negative, indicates a steeper regression line and greater impact. If the slope is zero, the regression line is horizontal, indicating no impact of the independent variable on the dependent variable.

The Y intercept usually has little interpretative value. To see why, note that another way to express it is as the value of Y when the X score is zero. Many measures in the behavioral and social sciences do not have a meaningful zero. Scoring a meaningful zero on a personality, attitude, or aptitude measure, for example, is rarely possible. Accordingly, the Y intercept rarely refers to an interpretable situation, though it does happen: Zero income, for example, is a perfectly meaningful, not to mention painful, situation. A final comment on the Y intercept concerns its sign. Note that a steep regression line may intercept the Y axis below the 0 point of the Y variable. In this case the Y intercept would have a negative value. This would be accurate but casts further doubt on the interpretability of the Y intercept in many if not most situations.

ANOVA Perspective on Regression

Earlier we noted that analysis of variance is a special case of regression. So it should come as no surprise that when we undertake a regression analysis using any respectable statistical program, the output includes an ANOVA summary table similar to the one we examined in Table 1.1. What does an ANOVA table look like when it is part of a regression analysis? Table 1.2 shows the summary table that results from the present regression analysis.

Remember that an analysis of variance splits up the total variability or differences in a dependent variable into components. Since the dependent variable of well-being has not changed, the bottom "total" rows in Tables 1.1 and 1.2 are identical. However, while the summary table in Table 1.1 showed component rows for between- and within-groups variability, the corresponding rows in Table 1.2 now refer to regression and residual sources. What exactly are these components? To answer this, we return to the "best-guess" way of thinking.

When we looked at well-being on its own, we said that the mean was our best guess for predicting individual scores, and the sum of squares (12) and

Table 1.2 ANOVA Summary Table Showing the Relationship Between Positive Affect and Well-Being

Source of Differences	*Sum of Squares*	*Degrees of Freedom*	*Variance*
Regression	8.44	1	8.44
Residual	3.56	8	0.45
Total	12.00	9	

variance (1.33) around this mean of 3 captured the extent of our failure to predict. But now we should be able to improve our predictions if it is true that positive affect accounts for differences in well-being. To see this in action, we can refer to Figure 1.6, which is a copy of Figure 1.5 with all cases except person J removed.

If we use the mean of 3 as a basis for predicting this person's score, the error or residual would be 3 – 1 = 2: the length of the vertical line between J and the dotted horizontal line. As we noted earlier, the regression equation does a relatively poor job of predicting this individual's well-being score: a predicted score of 2.26 versus an actual score of 1 and so an error or residual of 1.26. This may be relatively poor, but the key point is that the regression information has increased our predictive power, or conversely decreased the amount of predictive error for this individual. This case-by-case analysis is interesting, but how can we combine the information into hit-and-miss indicators for the whole sample?

As we just noted, the length of the vertical line from J to the mean represents the total error we make in predicting this individual's well-being score using the mean. The segment of this imaginary line between the mean and the regression line, labeled a, represents the gain in predictive power due to our using positive affect information, that is, the gain due to regression. The remaining segment between the regression line and the individual score, labeled b, represents the amount by which we are still failing to predict, that is, the residual. The segment due to regression can be calculated for each individual. These values can then be squared and summed to arrive at the sum of squares due to regression: 8.44 in Table 1.2. This has 1 degree of freedom (defined by the number of independent variables), so the regression variance is also 8.44. Similarly, the residual segments can be squared and summed across the cases and result in a residual sum of squares of 3.56. This sum of squares has 8 degrees of freedom (defined as the number of cases minus the

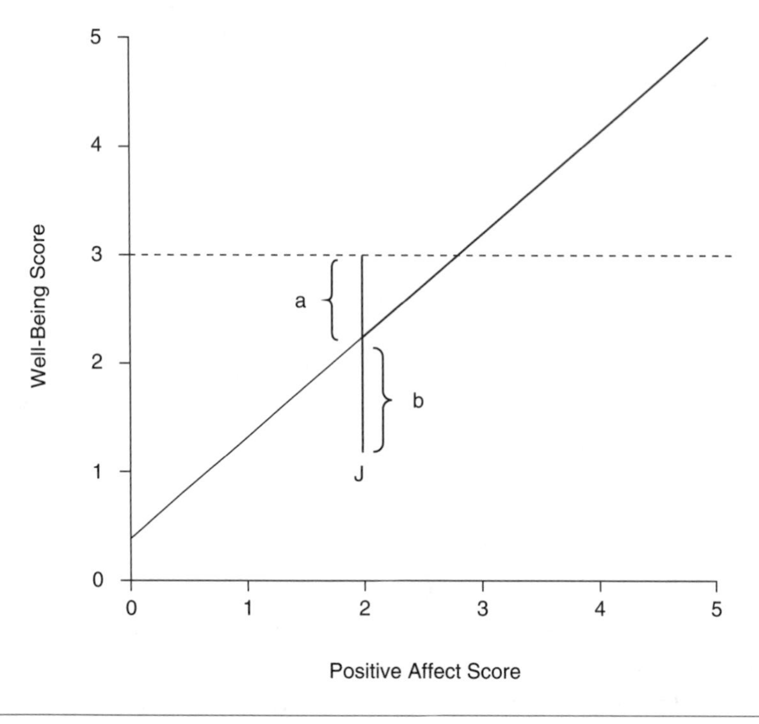

Figure 1.6 Scatter Plot Showing the Relationship Between Positive Affect and Well-Being Showing Only Case J

number of independent variables minus 1), and therefore the residual variance is 3.56/8 = .45. It is also worth noting that in regression we often refer to the square root of the residual variance, which is known as the **standard error of estimate.** In the present data this has a value of .667 and indicates how spread out the data points are around the regression line in the form of a standard deviation (the square root of a variance).

The analysis of variance in a regression context therefore partitions the total differences or variability in the dependent variable into two components. The regression component captures differences that can be accounted for by differences in the independent variable. The residual component captures the differences that remain unaccounted for. As before, if we express these components as sums of squares, the two components literally add up to the total, as do the degrees of freedom. Also as before, we can combine elements of the table to form indices of interest. Dividing the regression variance (or mean square) by the residual variance results in the F ratio: 8.44/.45 = 18.8 if we allow for rounding error. We noted earlier that the *F* ratio is not a direct measure of the strength of a relationship and has other uses that we will explore in Chapter 2 and beyond. It is noted here mainly because it is typically included in an ANOVA summary table.

If we wish to quantify the relationship between positive affect and well-being in terms of *individual differences,* we can divide the regression sum of squares by the total sum of squares: 8.44/12 = .703. This is a direct parallel to eta^2, which we encountered earlier, and is known as r^2 or, more formally, the **coefficient of determination.** This, too, can be multiplied by 100 and interpreted in terms of explained variability. So in these data positive affect explains 70.3% of the variability or individual differences in well-being. If we take the square root of the coefficient of determination, we produce the well-known **Pearson's correlation coefficient *r*,** which here has a value of .84. Unlike r^2, Pearson's *r* can be positive or negative and so can take on values between −1 and +1. Stronger relationships move *r* closer to +1 or −1, while a value of zero indicates the total absence of a relationship, as long as the relationship is best captured by a straight line. The value of .84 indicates a strong positive relationship such that higher scores on the positive affect measure are strongly associated with higher well-being scores.

It is important to reiterate the different types of information provided by regression and correlation coefficients, respectively. Regression coefficients, or more particularly the slope, indicate how an independent variable accounts for *group* differences in scores. The slope quantifies how group differences in the independent variable impact on group differences in the dependent variable. In contrast, correlation coefficients and their relatives show how far *individual* differences can be accounted for. Unlike regression coefficients, they express relationships in a symmetrical form. If we were to switch the independent and dependent variable status of positive affect and well-being, the regression coefficients would be different. Statistically speaking, the impact of positive affect on well-being is not the same as the impact of well-being on positive affect. However, this reversal would leave *r* and r^2 unchanged. We can accurately talk about each explaining 70.3% of variability in the other or about both sharing 70.3% of their variability. As a final demonstration of the difference between regression and correlation coefficients, it is worth revisiting Figure 1.5 and noting that the regression coefficients define the regression line itself, whereas *r* and r^2 convey how tightly the data points cluster around the line. A higher slope value (positive or negative) indicates a steeper line, whereas a higher correlation value (positive or negative) indicates a tighter distribution of data points around the line.

Inside the Coefficients

To complete this introduction to simple regression, we will look inside the regression and correlation coefficients, to demonstrate again how much we can achieve with just a few building blocks and to show how their values are calculated. Inside we will find the mean, variance, and standard deviation. But note that none of these actually captures the bivariate *relationship* between two sets of scores; each is a so-called univariate statistic. So we need to introduce one more building block called the **covariance,** which summarizes the strength and direction of a bivariate relationship.

As the name suggests, the covariance is a close relative of the variance. To compute the variance, we calculated a deviation score for each individual, squared and summed them across individuals to produce a sum of squares, and then took the average by dividing the sum of squares by the degrees of freedom. In the bivariate situation each individual has *two* deviation scores, one for well-being and one for positive affect in the present example. The key move in developing the covariance is to *multiply* each pair of deviation scores to form so-called cross-product scores. All of these scores are then added together to produce the **sum of cross-products.** This is a direct analog of the sum of squares, but whereas the sum of squares captures all of a single variable's variability, the sum of cross-products captures all of the *covariability* between two variables. This total covariability can then be averaged by dividing through by the degrees of freedom, which again is the number of individuals minus 1. The result is the covariance, which, for the well-being and positive affect relationship, is approximately 1. The sign of the covariance, positive in this case, indicates whether the relationship is positive or negative. However, the magnitude of the covariance is not readily interpretable as it stands. Instead it becomes the key element inside regression and correlation coefficients whose magnitude can be interpreted.

The slope can simply be defined in general as the covariance of X and Y divided by the variance of X. This definition highlights how the slope indexes the strength and direction of the X-Y relationship relative to differences in X, the independent variable. So for the present data, the slope would be $1/1.07 = .94$, as we noted earlier. This formula for the slope also reveals why the magnitude of the slope would change if we reversed the status of the independent and dependent variables. Although the covariance would not change, the variance in the denominator of the slope would now be that for well-being, which is 1.33. So the slope indexing the impact of well-being on positive affect would be 1/1.33 = .75. The slopes for the impact of X on Y and Y on X are the same only when the two variables have identical variances. With the slope defined in terms of a covariance and variance, we can now use it in a simple formula for the Y intercept. The Y intercept is just the product of the X mean and the slope subtracted from the Y mean. For the present data, this would be 3 – (2.8)(.94) = .37, as we noted earlier.

Finally, what are the components of Pearson's r correlation coefficient? Not surprisingly, the covariance again plays the key role as it captures the strength and direction of a bivariate relationship. To convert the covariance into a correlation, we simply divide it by the product of the X and Y standard deviations. So Pearson's r for the well-being and positive affect relationship would be 1/(1.16)(1.03) = .84, the figure we produced earlier by another route using the sums of squares from the ANOVA table. We also noted that r and r^2 capture a bivariate relationship in a symmetric fashion: Reversing the status of X and Y has no effect. This is now further apparent in the formula using the covariance and the standard deviations. The covariance of X and Y is the same as that for Y and X, and it does not matter in which order we multiply the standard deviations: a truly symmetric statistic.

Overview of Section 1.1

This completes a fairly lengthy introduction to the basic statistics used to quantify differences in one variable, and the relationship between differences in two variables, when the dependent variable is in the form of scores. Given this length, a brief overview may be helpful and serve to reiterate how these univariate and bivariate statistics are all derived from the same building blocks. In describing differences in one variable, we moved from the mean to the deviation score, to the sum of squares, to the variance via the degrees of freedom, and finally to the standard deviation. We then discussed how to discover whether category differences might account for these score differences, using an approach called the analysis of variance. This involved dividing up the total variability in the scores into between- and within-group components using sums of squares, degrees of freedom, and variances. From there we derived three statistics: the F ratio, eta^2, and Wilks's lambda, all using ratios of the statistics in the ANOVA summary table. However, we noted that the difference between mean scores on the dependent variable across categories remains the most fundamental way of describing the effect of the categories on the scores at the *group* level of analysis.

We then turned to simple regression, which is used to discover whether score differences in an independent variable might account for score differences in a dependent variable. The regression line, identified by its slope and Y intercept, was introduced as a way of capturing the impact of the independent variable on the dependent variable at the group level. We then saw how analysis of variance could be used to divide up the total variability in the dependent variable into predictable (regression) or unpredictable (residual) components. Again we found that elements in the ANOVA summary table could be combined to form further statistics: the F ratio, the coefficient of determination (r^2), and Pearson's correlation coefficient (r). Finally, we introduced the building block of the covariance and showed how it, in conjunction with means, variances, and standard deviations, is used to build the bivariate regression and correlation coefficients.

The mean, sum of squares, degrees of freedom, variance, covariance, and standard deviation make up an immensely powerful building set. In the foregoing sections we have used them to build a first story of techniques for analyzing bivariate data, revolving around the analysis of variance and simple regression. In Part 2 of the book we will move to the higher level of multivariate analysis and see, for example, how simple regression and correlation can be extended into multiple regression (Chapter 4) and factor analysis (Chapter 7). But however complex the situation becomes, we will continue to combine these basic building blocks.

1.2 ANALYZING DATA IN THE FORM OF CATEGORIES

To complete this first chapter, we shift our attention to data that are all in the form of categories. Imagine that we ask a group of 10 men and a group of 10 women the simple question of whether or not they are generally happy, with the intention of finding out whether there is a gender difference. This is the same research question we asked earlier when exploring analysis of variance, but now the data have a different form. The present data come from two

Table 1.3 Contingency Table Showing the Relationship Between Gender and Happiness

	Women	*Men*	*Total*
Happy	ABCDEFGH 8	IJKL 4	12
Not Happy	MN 2	OPQRST 6	8
Total	10	10	20

categorical variables each with two categories: men/women and happy/not happy. Since the measurement procedures here do not involve rank ordering or assuming equal distances between categories, all we can really do is to count the number of individuals within and across categories and develop summary statistics from there. Accordingly, this section will be considerably shorter than Section 1.1. However, when we subsequently move to multivariate analysis of categorical data in Chapters 5 and 8, it will again be apparent how powerful techniques can be developed from very simple beginnings.

Table 1.3 shows the 20 individuals, labeled A–T, distributed on a set of three ladders, each with two rungs. The left-hand ladder shows the happiness distribution for the women, and the middle ladder the distribution for the men. As noted above, all we can do with categorical data is to count frequencies, and these are shown below the letters on each rung. The right-hand ladder shows the total happiness distribution just in the form of frequencies to avoid repetition and clutter: 12 happy individuals (A–L) and 8 not happy ones (M–T). Note that, since neither variable has ordered categories, the order of rungs and of category ladders is arbitrary. This type of display is known as a **contingency table** since it shows how the distribution of a dependent variable (happiness) is contingent on the distribution of an independent variable (gender). What have been referred to as "rungs" are more conventionally known as cells. The frequencies in the four cells within the table are called the **conditional frequencies,** and those around the edges that show the total distributions for the two variables are called the **marginal frequencies.** The bottom right-hand cell contains the total frequency, which of course is equal to the number of individuals.

Given the limited form of the data, how might we quantify the distribution of a single variable such as the dependent happiness variable, shown in the right-hand column? Neither the mean nor the median would be a meaningful summary of the middle of the distribution. Instead we use the **mode,** which is simply the highest frequency, that is, 12. So the modal category is the happy one; the most common status for this group of people is to be happy. To capture differences or variability, we can make use of ratios of frequencies in two ways. Dividing a marginal cell frequency by the total and multiplying it by 100 provides a difference statistic in the form of a percentage. So 100(12/20) = 60% of the group are happy, while 100(8/20) = 40% are not. These relative frequencies are often treated as *probabilities,* for example, a case has a 60% chance of being happy. Alternatively, we can form a ratio of marginal cell frequencies to produce an **odds** statistic. The odds of being happy in this group are 12/8 = 1.5; conversely, the odds of not being happy are 8/12 = .67. Notice that an odds of 1 indicates an equal split and therefore maximum variability, as in the case of the gender variable. The further the odds deviate above or below 1, the less the variability or difference on that variable.

One reason for introducing the odds statistic is that it can easily be developed into a useful bivariate statistic that captures the relationship between two categorical variables. If we look inside the table, we can see that the so-called **conditional odds** of being happy for women are 8/2 = 4, and the conditional odds for men are 4/6 = .67. If we divide one of these odds by the other, we arrive at the **odds ratio.** This will be 4/.67 = 5.97, which says that women are nearly 6 times as likely to be happy than not, compared with men in this group. This method of capturing the relationship between categorical differences has become a commonplace of public health messages. Statements such as "smokers are twice as likely to suffer a heart attack as nonsmokers" are based on the calculation of odds ratios. Unlike regression and correlation coefficients, where the absence of a relationship is indicated by a value of zero, an absent relationship between two categorical variables results in an odds ratio of 1. So the further the odds ratio deviates above or below 1, the stronger the relationship.

In Subsection 1.1.2, on analysis of variance, we briefly encountered a statistic called the *F* ratio. This was crudely described as a way of indexing the extent to which group differences on a dependent variable are greater than what might be expected on the basis of individual differences. We can take a similar approach with wholly categorical data, using a statistic called **chi^2.** This statistic can be used to quantify how much greater the group differences we see in the conditional frequencies in Table 1.3 are than we would expect given the individual differences evident in the marginal frequencies. Notice that this is equivalent to asking how far the odds ratio differs from 1, or to asking whether there is a relationship between gender and happiness in these data. All other things being equal, the value of chi^2 increases with the strength of the relationship, and in the present case it has an approximate value of 3.33. However, as with the *F* ratio, chi^2 is not a direct measure of the strength of a relationship. Its true function will become clearer in Chapter 2 as a so-called test statistic. For now, it is sufficient to note that it can be transformed to arrive at a correlation coefficient called the **phi coefficient.** The transformation involves dividing chi^2 by the total frequency and then taking the square root: The square root of 3.33/20 = .41. This turns out to be a special version of Pearson's *r,* which can be used to quantify the relationship between two dichotomous variables, that is, each having two categories. Accordingly, we can report that gender and happiness are correlated .41 in the present group, or we can square phi and multiply it by 100, as we did for r^2, and say that gender explains about 16.8% of the variability in happiness.

Much more could be said about the nature of chi^2, the analysis of contingency tables with more than 4 cells, and other types of correlation coefficients for categorical data. However, the ideas and statistics we have briefly discussed are sufficient as a foundation for later chapters. As noted earlier, we will build directly on this foundation in Chapter 5 when we encounter logistic regression, and especially in Chapter 8 when we explore log-linear analysis in general to see how it can be used to analyze contingency tables with more than two categorical variables.

1.3 FURTHER READING

Good introductions to basic data analysis can be found in Rowntree (2003) and Rosnow and Rosenthal (2001). More extensive treatments of univariate and bivariate analysis techniques can be found in Hays (1994) and Rosenthal and Rosnow (1991). Good, focused introductions to ANOVA and simple regression are available in Keppel, Saufley, and Tokunaga (1993) and Darlington (1990), respectively.

TWO

DECIDING WHETHER DIFFERENCES ARE TRUSTWORTHY

The purpose of any statistical analysis is to provide an accurate and informative answer to one or more questions. The analyses in Chapter 1 were obviously contrived and intended to elucidate the analytic procedures themselves. In a real-world context, though, these types of analyses are used to reach conclusions that have theoretical or practical consequences. But how much trust can we place in the results of even simple statistical analyses? What sorts of concerns should be routinely reviewed before we take statistical results seriously? This issue of trustworthiness is so important that we will review it each time we encounter a new multivariate technique in Part 2. In this chapter we introduce the main concerns that apply to even simple analyses and that subsequently recur in a variety of forms in more complex multivariate analyses.

Issues of trustworthiness can be organized into four categories relating to sampling, measurement, the role of chance, and the legitimacy of adopting a particular statistical technique. "Trustworthiness" is not a conventional term in quantitative data analysis, but it is a useful word for encompassing this wide range of issues. In Section 2.1 on sampling we will examine in general terms how the number of cases in an analysis, their attributes, and their mode of selection influence the trustworthiness of that analysis. This discussion will treat a case as an individual person, but the issues remain the same when cases are other entities such as schools, communities, or business organizations. Statistical analyses require that attributes of cases have been measured using at least a categorical scale. It is the measurement procedures that generate the numbers from which the statistical analysis seeks to extract patterns. Thus in Section 2.2 on measurement we will review the different types of measurement scale and how they relate to different types of analysis. Then we will discuss how deficiencies in the consistency (reliability) and accuracy (validity) of measurements can influence the trustworthiness of analyses.

Section 2.3 revolves around the question of how far a particular statistical result may be due to chance. There are various ways to approach this question, but the most popular in the behavioral and social sciences is a procedure called null hypothesis testing. In its most common form this procedure makes use of so-called test statistics, such as F and chi^2, to evaluate whether an observed difference or relationship is different from zero. Concluding that the result is different from zero is seen as providing evidence that the result is unlikely to be due to chance. Since null hypothesis testing is so ingrained in statistical analyses conducted by social scientists and is the source of some contention, we will devote considerable attention to it in Section 2.3. This section will also introduce an alternative approach to evaluating the role of chance called estimation. This strategy makes use of many of the same tools found in null hypothesis testing but avoids some of the pitfalls of the latter that critics have identified. Both null hypothesis testing and estimation are found throughout multivariate analysis, so it is important to acquire a firm grasp of their logic and their strengths and limitations before we move into Part 2.

In Chapter 1 we noted that even a simple summary statistic such as the mean requires certain conditions (interval scaling and normality) for it to be effective and therefore trustworthy. The addition of test statistics to our analytic tool kit brings more conditions to be met, and these are usually referred to as statistical assumptions. In Section 2.4 we will review the statistical assumptions that have to be met in simple analyses for the results to be regarded as trustworthy. As usual, these assumptions will turn out to be foundational in the sense that they constantly reappear in more complex analyses in slightly different forms. So again, grasping these core ideas now will make the subsequent task of making sense of multivariate techniques much easier.

Before we embark on the details of trustworthiness, two other broad issues deserve brief comment. The notion of trustworthiness has been introduced as if it first arises once an analysis is complete. However, in practice it is a concern throughout the research process. Selecting an appropriate sample, adopting reliable and valid measuring instruments, and choosing legitimate statistical techniques are crucially important activities in planning an effective research project. Then, once data have been collected, issues of trustworthiness should be reviewed again before statistical analyses are undertaken. So the issues we are about to explore are pertinent at all stages of a research project. The only reason that the discussion appears to emphasize the postanalysis stage at the expense of others is the book's primary objective of helping readers to understand and critically evaluate completed research.

The four issues discussed in this chapter do not exhaust all aspects of trustworthiness. At the beginning of this chapter it was suggested that the purpose of any statistical analysis is to provide an accurate and informative answer to one or more questions. Even if all of the concerns we will discuss in the following sections were dealt with satisfactorily for a particular analysis, it still would not guarantee an accurate and informative answer. Such an outcome requires also that research questions, theories, and, where appropriate, hypotheses have been well specified and that an appropriate overall design strategy has been adopted. For example, the results of a well-designed experiment have a very different status from those of a survey, especially when causal questions are being asked. Even for highly accurate experimental results, the absence of a good theoretical framework may render them uninterpretable and therefore uninformative. These are important and fundamental aspects of trustworthiness that require discussion if we are to appreciate the limitations of statistical and especially multivariate techniques. But such a discussion is better postponed to the end of Chapter 3 as a final preliminary before we move into Part 2.

2.1 SAMPLING ISSUES

The trustworthiness of statistical results depends partly on the number of cases in the analysis, their attributes, and the way in which they were selected. Jointly these factors can be loosely referred to as sampling issues. As a general rule, trustworthiness increases with the number of cases for several reasons. As we have seen, the outcomes of analyses take the form of summary statistics such as mean differences and regression slopes. These descriptive statistics generally become more stable and reliable as the number of cases increases. If the cases in the analysis are being treated as a sample, having more cases also increases the trustworthiness of the inferences that can be drawn about the population from which the sample was drawn. In Section 2.3 we will reanalyze the data from Chapter 1 from this perspective of statistical inference. We will conclude, for example, that our finding of a gender difference in well-being in favor of women is not a trustworthy basis for inferences beyond this sample. From this analysis we cannot reliably conclude that women in the population sampled enjoy better well-being than men or that they do not. As we will see, this indeterminate conclusion is very much tied up with the inadequate sample size used in the analysis.

When we are considering the number of cases, it is important to distinguish between the number that were acquired in a study and the number that were actually used in a given analysis. The latter may be considerably smaller than the former, usually because of the ubiquitous problem of missing data. It is a rare study in which every participant provides usable data on all of the variables. Participants may forget or choose not to provide answers to certain questions or may answer in such a haphazard or biased fashion that the data are not trustworthy. In most situations an analysis can be carried out only on those participants who have provided usable data on all of the variables in that analysis. Accordingly, the analyzable sample size may shrink alarmingly even where each participant is responsible for only a little missing data. This problem becomes particularly worrisome in multivariate analyses in which the whole point of the exercise can become undermined as it may not be possible to examine all of the variables at once because of inadequate sample size. For this reason, methods have been developed to "impute" or replace missing data with unbiased estimates based on the participant's other responses or on those of other participants. These **imputation methods** can help to counter the missing data problem, but they inevitably carry other costs in terms of the trustworthiness of results. Thus it is always wise to check whether imputed data have been used in any analysis.

More cases in an analysis also allow the analyst to address another issue of trustworthiness—that of the generalizability of results. As we will see in later chapters, multivariate analyses provide the rosiest pictures of data. In Stevens's colorful words: " . . . there is great opportunity for capitalization on chance, seizing on the properties of the sample" (Stevens, 2002, p. xiii). It is always important, therefore, to ask how well a particular result may replicate. All sciences place great store on replication of findings across different studies. But given the practical and theoretical difficulties of strict replication, it is helpful and reassuring to attempt replication within the same study. One way to achieve this is by acquiring sufficient cases so that, for example, half of the cases can be treated as a **holdout sample.** As the name implies, these cases are not used in the main analysis but subsequently form a replication or cross-validation sample on which the generalizability of the results from the main analysis can be tested. If this strategy is followed, the sample size requirement thereby doubles.

Two other general points may help us to tighten our grasp further on the sample size issue. The first is that the smaller the differences or relationships that are being sought, known generally as the **effect size,** the more cases are needed to detect them. Subtle effects may be of great interest, especially in theory-driven research, but their presence may be missed if too few cases are analyzed. Conversely, relatively few cases will be needed to detect a large effect, and time and resources may be wasted in recruiting more participants than are necessary. The second point is that, in general, the more variables that are being analyzed at once, and the more complex their relationships, the more cases are needed. This is why, as we will see in later chapters, a common way to address the question of sample size is to recommend a minimum ratio of cases to number of variables.

So far we have reviewed a number of reasons why the number of cases in a study should be larger rather than smaller, but we have yet to confront the question of how large is large enough. The answer to this question varies according to the data analysis technique, so we will consider it in each of the chapters in Part 2. For now, a very rough guide would be that an absolute minimum of 50 cases is needed for any single multivariate analysis, but that in practice one or more hundreds of cases are typically required for trustworthy results. Add to this the concerns about missing data, replication, size of effects, number of variables, and complexity of relationships, and it becomes apparent that multivariate techniques are hungry for cases. Multivariate techniques provide a very powerful way of analyzing complex data, but the price is an investment in relatively large sets of participants. This may be simply impracticable in many situations, in which case the benefits of multivariate analysis have to be forgone.

The trustworthiness of results depends not just on how many cases there are, but also on who they are. The broadest requirement is that the participants be appropriate for the research question and design. If we are asking questions about well-being, to whom should the answers apply? For example, do we have in mind the general population in reasonable mental health, or do we want to include individuals with clinical problems? The scope of the research question and its underlying theory and assumptions should be critical determinants of what constitutes an appropriate sample. Unfortunately, this linkage is not always explicit, and so issues about how to select the sample, and particularly what cases should be excluded from analyses, may not be treated rigorously, if they are raised at all. Other research design issues are also pertinent here. For example, if the study is an experiment in which the objective is to enhance well-being, there is no point in including individuals whose state is unchangeable because of clinical problems or medication. Broadly speaking, the participants in a study should be exemplars of those people to whom the research question applies and in whom the phenomena of interest may be found or induced. More specifically, participants should be able and willing to provide the information or to undertake the activities required. Many studies are quite demanding of participants' knowledge, memory, processing of information, powers of expression, and willingness to disclose. Accordingly, it is always pertinent to ask about the match between a study's requirements and the attributes of the participants. If the match is poor, the trustworthiness of any results, however well analyzed, is suspect.

The final sampling issue relating to trustworthiness concerns the way in which the participants in a study have been selected. There are many ways to select a sample, but within these there is a major distinction between **probability and nonprobability sampling strategies.** In the former, sample members are chosen from a fully defined population in such a way that the probability of each member's selection has a known value. For the most straightforward version of this strategy—the **simple random sample**—each member of a population has the *same* probability of being selected as a sample member. But there are many more complex probability strategies underpinned by sophisticated statistical theories of probability. In contrast, nonprobability samples are assembled without formal reference to a defined population and without a selection strategy that generates known probabilities. Instead participants are chosen using other "purposive" or appropriateness criteria such as those reviewed above.

This distinction may tempt us to treat probability samples as the gold standard and to condemn nonprobability samples as less than trustworthy. But this seems to be at odds with the undeniable fact that most samples in the behavioral and social sciences are not probability samples. Given the complexity of the issues here and space limitations, only a few general, but hopefully reassuring, comments are possible. The most important point is to reflect on the link between research questions and sampling strategies. Probabilistic samples are required if the objective is to estimate with precision and accuracy the prevalence or strength of a phenomenon in a defined population. It is not surprising, therefore, to find national attitude surveys being used as common examples when probability sampling is introduced. However, the objective of a great deal of behavioral and social research is to discover whether a phenomenon occurs at all and, if so, the nature of its causes and consequences and generally how it works. Theories and hypotheses in these disciplines rarely specify or predict the prevalence or strength of a phenomenon, and so there is no need to worry about the question of whether samples provide quantitative pictures that are representative of populations.

A further issue concerns the very definition of populations. Probability sampling works only if there is a defined population from which samples are drawn. But it is often unclear to what populations a given theory, hypothesis, or research question should apply. It seems strange to invest considerable energy in assembling a representative sample of a population, such as the student roll in a particular university, unless the choice of population can be shown to be of some theoretical or practical significance. At a more pragmatic level, it has also been pointed out that attempts to acquire a representative probability sample often founder because of nonresponse. So even if a true probability sample is dictated by the objectives of a study, it may be practicably unobtainable.

Although a probability sample may not be necessary or practicable, its absence does place constraints on the interpretation of statistical analyses. In Sections 2.3 and 2.4 we will explore the use of statistical inference, particularly null hypothesis testing, as a way of deciding whether statistical results are due to chance. There again we will encounter the notions of samples and populations, and it might seem that the sampling and inference strategies map nicely onto each other. However, this is not the case. If a probability sample has been used, it is appropriate to think in terms of making inferences from the actual study sample to the actual population. Under these circumstances, applying precise probabilities to outcomes and estimating what might happen if another sample were drawn from the same population are legitimate moves. But if a nonprobability sample has been used, the interpretation of chance, probabilities of outcomes, and replicability become much less well defined. All of this means that evaluating the extent to which chance may account for statistical results is considerably more constrained and hazardous in the absence of a probability sample. Nonprobability sampling does not invalidate the findings of a study, but it does put limits on how far chance can be eliminated as a source of untrustworthiness.

2.2 MEASUREMENT ISSUES

However sophisticated a statistical analysis becomes, the results cannot be trustworthy if the data have been generated by inappropriate or deficient measurement procedures. In this section we will briefly unpack some meanings of "inappropriate" and "deficient," which bear on the relationship between measurement and statistical analysis. First, in Subsection 2.2.1 we will review the notion of a **measurement scale**—a set of rules for assigning numbers. Then, in Subsection 2.2.2 we will reflect on how measurements that are inconsistent and/or inaccurate might distort the outcomes of statistical analyses.

2.2.1 Measurement Scales

There are many ways to assign numbers to or "scale" phenomena. A conventional way of organizing the most commonly used is to classify them as categorical (also known as nominal), ordinal, interval, or ratio measurement scales. We can illustrate the meaning of each scale by imagining different ways of measuring happiness. A **categorical scale** involves assigning each case to one of a set of categories, such as "happy" and "not happy." The categories must be mutually exclusive and comprehensive, so that every case can occupy an unequivocal category. Strictly speaking, this is not measurement as such since no numbers are being assigned, only names (hence "nominal"). But it does make a useful foundation on which to build. An **ordinal scale** has all of the features of a categorical scale but adds the requirement that the categories can be meaningfully ordered. If this is true, then cases can be assigned to ranks such as 3: "very happy"; 2: "moderately happy"; and 1: "not at all happy." Numbers have now appeared, but it is important to appreciate that they denote only an ordering of categories. An **interval scale,** as we saw in Chapter 1, adds the further requirement that the distance between each ordered category is the same. There we used a 5-point well-being scale and made this equal interval assumption. Finally, a **ratio scale** adds the requirement of a meaningful zero to anchor the scale. The key word here is "meaningful." We could devise a happiness measure with a zero point, but it would be defensible only if theory and data showed that zero happiness was conceptually sound and actually occurred in coherent ways.

Making decisions about appropriate measurement scales can (and should) lead the researcher to question and refine the nature of a phenomenon such as happiness. Once the decision has been made, the choice has consequences for how the data generated might legitimately be analyzed. So the measurement scale must be appropriate both in terms of theory and of data analysis. The latter implication stems from the fact that different measurement scales permit different mathematical operations. As we noted in Chapter 1, for example, analyses involving the mean and its derivatives require data from measures with at least an interval scale. Each data analysis technique, including multivariate ones, has scaling requirements for the variables in the analysis. We will find that factor analysis (Chapter 7), for example, requires interval level data, while log-linear analysis (Chapter 8) requires categorical data. So measurement scales are relevant to the trustworthiness of results because they partially determine what types of analysis may legitimately be used.

This discussion makes the issue of measurement scales sound very daunting and constraining. Two observations may help to soften this impression. First, it is helpful to think of a measurement scale more as a working assumption about how a measuring instrument works than as an intrinsic property that can be clearly demonstrated. As a common example, the equal interval assumption is one that many analysts are prepared to make about rating scales that could be regarded as ordinal because it makes theoretical sense and produces data that are coherent. Moreover, duplicating the analysis with techniques that require only ordinal assumptions rarely changes the message of the analysis. This is not an argument for "anything goes" but an attempt to reduce exaggerated concerns about the "real" nature of a scale.

Second, we will find that many techniques can be adjusted to accommodate variables with different scaling properties. For example, although multiple regression (Chapter 4) is designed to analyze interval level variables, categorical and ordinal level variables can be included (as independent variables) through a technique called dummy coding. So the relationship between measurement scales and analysis technique is not as rigid as it first appears. This is not only reassuring, but also opens up possibilities for parallel analyses of the same data using different techniques. This potential for shining different lights on data from different directions can only enhance the analyst's ability to detect patterns of interest.

2.2.2 Measurement Quality

The trustworthiness of statistical results can also be undermined if the data have been gathered with deficient measuring instruments. The quality of a measurement procedure is conventionally evaluated in terms of whether the procedure produces data that are reliable and valid. **Reliability** of measurement refers to how far the data are contaminated with random errors that make them *inconsistent.* **Validity** of measurement refers to how far the data are subject to systematic errors or bias that makes them *inaccurate.* These abstract notions can be made more concrete with an example of an actual measuring instrument that we might consider adopting to assess positive affect.

Watson, Clark, and Tellegen (1988) have devised a mood measure called the Positive and Negative Affect Schedule (PANAS). Respondents rate each of 20 mood adjectives on a 5-point scale ranging from "very slightly or not at all" to "extremely." Instructions can be adjusted according to the time frame of interest, ranging from current mood, through days, weeks, and years, to mood in general. Ten of the ratings can be summed to produce a negative affect score, and the other 10 to produce a positive affect score. Our interest is only in the positive affect part of the measure.

In their 1988 article Watson et al. provide extensive information in support of the reliability and validity of the two PANAS subscales. One common way of evaluating reliability is to estimate how well the elements of a measure operate in concert: the measure's **internal consistency.** This can be quantified by correlating the item scores for a sample of respondents and calculating a statistic called **Cronbach's alpha.** It is usually suggested that an alpha of .7 provides minimal reassurance of internal consistency, though a higher value is desirable. Watson et al. report positive affect alphas for six samples that range between .86 and .90—a very reassuring picture. It is also possible to evaluate a measure's consistency *over time* by correlating scores from a group of respondents who are tested and then retested at a later date. The **test-retest reliability** correlations for the positive affect measure are reported in the range .47 to .68, depending on the time frame of the instructions. Assuming that some of the inconsistency over time is due to actual mood change, these are again reassuring results.

A measure may produce consistent results, but they may be consistently inaccurate. Different sorts of evidence are needed to demonstrate the validity or accuracy of a measure. The **content validity** of a measure, as the name suggests, is reflected in a correspondence between its content and that of the phenomenon or construct that is to be measured. In the present example we would need to ask whether the 10 positive affect adjectives capture the construct of positive affect as we want to define it for our research. Do the adjectives cover all of the important facets? Do they exclude facets of other similar constructs? Validity can also be assessed statistically by examining the pattern of correlations of scores with those from other measures. Watson et al. report, for example, that scores from the positive affect subscale correlate negatively in the range −.19 to −.36 with scores from various established measures of distress and dysfunction. These figures provide some reassurance about the so-called **construct validity** of the measure.

If data that are deficient in reliability or validity are entered into a statistical analysis, what might the consequences be? The random errors produced by unreliable measurement can lead to underestimates of the magnitude of actual differences or relationships, or to difficulties in assessing whether they are due to chance. Unreliability becomes even more of a problem in multivariate analyses. In this context we often want to examine a relationship between two variables while controlling the influence of other variables. The problem is that if these other variables—called **covariates**—have been measured unreliably, all sorts of unpredictable distortions may percolate through the analysis. Darlington (1990) provides a helpful discussion of this issue in the context of multiple regression and concludes: "The effect of measurement error in covariates may be the most important single weakness in regression analysis" (p. 204).

The statistical effects of data generated by invalid measures are harder to specify. For example, the means of such variables may be systematically under- or overestimated. But this may or may not contribute to distortions of differences and relationships among variables. The fundamental problem remains, though, that if the data do not represent the target variable accurately, conclusions about that variable are likely to be undermined. If positive affect scores do not accurately capture differences in positive affect, no manipulations in their statistical analysis will remedy the problem. It is worth adding in conclusion that this issue again takes on an even more serious aspect when we engage in multivariate analysis. As we noted earlier, much of this type of analysis involves the control of covariates—variables that might distort a relationship of interest between a particular independent variable and a dependent variable. As we will see in Chapter 3, this form of statistical control works by adjusting variables so that their nature may be changed. If their character is already ill defined or distorted because of measurement validity problems, subsequent adjustment can only compound the problem.

It should be clear from this discussion that measurement issues are important determinants of the trustworthiness of a statistical analysis. Moreover, we have seen various ways in which these issues can become particularly problematic in a multivariate context. Recent developments in statistical methods are providing sophisticated solutions to some of these problems. However, none of these solutions should be seen as grounds for not ensuring the appropriateness and quality of data before statistical analyses are undertaken.

2.3 THE ROLE OF CHANCE

In Subsection 1.1.2 of Chapter 1 we analyzed a small set of artificial data to find out whether women are more likely than men to experience higher levels of well-being. This analysis suggested that, as a group, women had a mean happiness score .8 higher than that for men. At the individual level, the analysis of variance showed that gender explained 13.3% of well-being differences. Although contrived, these data are consistent with research findings in general on this topic. But if this were a real study, how would we evaluate whether or not this particular outcome was simply due to chance and therefore trustworthy as evidence in support of a gender difference?

2.3.1 Evaluating Chance With Null Hypothesis Testing

In the social sciences the most common response to this question is to engage in a procedure called null hypothesis testing (NHT). In its most usual form this procedure involves three strategic moves. The first move in the present example is to hypothesize or imagine that in fact there is no gender difference and so any observed difference in a particular study must be only apparent and due to chance. This is the null hypothesis—in a sense the worst-case scenario that envisages a world where the suggestion of a gender difference is completely erroneous. The second move is to use statistical theory to calculate the probability of finding a difference as large as that recorded in our study, if the null hypothesis were true. Clearly, if the difference were large and its probability correspondingly small, then we should start to entertain doubts about the truth of the null hypothesis. The third move is to decide precisely how small that probability should be before we can actually reject the null hypothesis. So ultimately, the testing outcome depends on a probability "threshold": the so-called significance level.

It should be apparent from even this brief outline that null hypothesis testing is highly abstract, involving leaps of the imagination and the application of statistical theory. This degree of abstraction, and the counterintuitive nature of some of the procedures, can easily lead to misunderstandings and misapplications. In the following we try to avoid these pitfalls by exploring the logic of each of the three strategic moves and by applying the NHT procedure to all of the analyses we conducted in Chapter 1. In doing this we will be introducing technical terminology, so that more advanced treatments will subsequently be accessible. But, as usual, we will not be delving into the statistical foundations as such. The NHT procedure is contentious, and indeed all three of the strategic moves are problematic in some ways. However, for the sake of clarity, it will be helpful to provide an uncritical account in this chapter and to postpone any critical reflections until the final section in Chapter 3.

The Null Hypothesis

Null hypothesis testing is a form of **statistical inference:** a way of reaching out beyond the available data. So we first need to distinguish between the available data and other imaginary data that could be collected in principle. The data we actually have at our disposal are referred to as a **sample,** which, as we have seen, can be summarized with statistics such as means, mean differences, standard deviations, correlation and regression coefficients, and odds ratios. (In fact, strictly speaking, the term "statistics" refers to descriptors of sample data.) As the name implies, we imagine that the sample data have been drawn from a **population.** This is an imaginary set of data of which the sample is but one subset. Characteristics of the population can also be described with summary numbers that are called **parameters.** Each sample statistic has a corresponding population parameter. So, in our example the mean gender difference in well-being is a known statistic in the sample and an unknown parameter in the population. The general objective of statistical inference is to reach conclusions about unknown parameter values using known statistics values. In the present example we are using a known mean difference from a sample to reach conclusions about the unknown mean difference in the population, but the same inferential procedures can be used for any statistic and its corresponding parameter.

In order to apply statistical theory to research questions, it is necessary to recast them in precise numerical terms. Earlier we noted that usually the first move in null hypothesis testing is to hypothesize that the observed difference is due to chance factors in our particular sample and that in fact there is no difference. We can now formulate the **null hypothesis** more precisely by stating that the parameter, in this case the mean gender difference in the population, is precisely zero. This is the null hypothesis that is to be formally tested and ultimately accepted or rejected.

Three aspects of the null hypothesis notion are worthy of comment before we proceed further. First, note that formal null hypotheses refer to parameter values. Sometimes descriptions of research make it sound as though hypotheses refer to sample values. But since these are known, there is no need for hypotheses or conjectures about their possible value. This may sound like an academic point, but an initial appreciation that statistical hypotheses are about parameter values can help to avoid confusion later. Second, the statistical requirement in formulating a null hypothesis is that the appropriate parameter value be set to a precise numerical value. Usually this value is set to zero, partly because it represents the worst-case scenario, but also because it is often difficult or impossible to choose a nonzero value that fits the research context. But in principle, null hypothesis testing can be used when a parameter is set to a nonzero value if such a meaningful value is available. To make this point clearer, Cohen (1994) refers to the zero version of the null hypothesis as the "nil" hypothesis. Third, since null hypotheses refer to parameter values, it is crucial that the analysis focuses on the *appropriate* statistic and parameter. In our example, the mean gender difference in well-being is the focus of interest because it represents the effect of gender on well-being. In other analyses the focus will shift so that the chosen statistic and parameter always capture the research question of interest. This issue will become particularly salient when we move into multivariate analysis in which multiple hypotheses will be under consideration. To maintain clarity, it will always be important to ask ourselves which particular null hypothesis, statistic, and parameter are the focus of the analysis at any given point.

We have initiated the NHT procedure by formulating a null hypothesis that sets a parameter value to a precise numerical value. In the present case we have hypothesized that the population value of the mean gender difference in well-being is exactly zero. If this is true, then our sample difference of .8 in favor of women is a random blip and not a reflection of the true state of affairs. Our goal is to decide whether to accept or reject this particular null hypothesis, that is, to test it. If after the test the null hypothesis is accepted, the situation is well defined; but if it is rejected, how should that state of affairs be characterized? Null hypothesis testing is in fact constructed around a *pair* of hypotheses. Attention focuses on the null scenario, but it is also necessary to specify its partner—the **alternate or research hypothesis.** As the name suggests, the alternate hypothesis specifies alternate values of the parameter in such a way that the null and alternate values are mutually exclusive and jointly cover all possible values. This sounds complicated but usually just involves setting the alternate parameter values to nonzero. So, under the null scenario the mean gender difference in the population is zero, and in the alternate scenario, it isn't. If the test indicates that the null hypothesis should be rejected, then the alternate hypothesis is adopted by default. The word "default" is important here because it highlights the fact that only the null hypothesis is actually formally tested; the ultimate status of the alternate hypothesis emerges as a consequence of the null hypothesis test.

The version of the alternate hypothesis we have just encountered is one of two ways of framing it. This version is referred to as the **nondirectional, or two-tailed, form of the alternate hypothesis.** The former name is more informative for a basic understanding because it indicates an outcome in which, on average, women report a different level of well-being compared with men, rather than specifying a higher or a lower level. In other words, the direction of the gender difference is not specified; the mean difference is simply not zero. As you might expect, the other version is known as the **directional, or one-tailed, form of the alternate hypothesis.** In this version we are free to hypothesize, not just a difference, but also the direction of that difference. Since from the outset we have been suggesting that women may be happier than men, it might seem odd that we now opt for the nondirectional form of the alternate hypothesis. The reasons for this are complex and beyond the scope of a conceptual introduction to NHT procedures. But pragmatically we can note that almost all of the NHT analyses provided by statistical packages assume that a nondirectional alternate hypothesis has been adopted and provide results accordingly. Further, if we did decide in a particular situation to choose a directional alternate, it is simple to adjust the statistical results accordingly. In summary then, we have now completed the first move of the NHT procedure by specifying a null hypothesis that the mean gender difference is zero in the population and an alternate hypothesis that the difference is nonzero.

Finding the Conditional Probability

At this point in the proceedings we have an observed mean difference of .8 and a suggestion that this may be entirely due to chance, couched in the form of a hypothesis that the actual mean difference is zero. The next move is to find the probability of this difference value (or a larger one) occurring in a particular sample if the null hypothesis were true. Put more technically, we want to know the conditional probability of the sample value under the null hypothesis. Since all we have at our disposal are our sample statistics, we again need to exercise our imaginations and draw on some statistical theory.

First we imagine that the sample mean difference of .8 is just one of an infinite set of sample differences that we could have observed, each based on 10 individuals. Were we to draw other samples of the same size from this infinite set, the difference values would vary. If the null hypothesis is true such that the actual mean difference is zero, we would expect most sample values to be zero or close to it; nonzero values would be less and less common the more they deviated from zero. In other words, if the null hypothesis is true, there is a high probability that a given sample difference would be close to zero and decreasingly lower probabilities for differences that are further and further from zero. Note that by adopting a nondirectional alternate hypothesis earlier, we have decided to ignore the question of whether sample differences are greater or less than zero. Our only interest is in how far a sample difference deviates from zero, so whether we view our particular sample difference as +.8 or −.8 is irrelevant.

Remember that what we want to know is the probability of a sample mean difference of .8 occurring, assuming that the actual difference in the population is zero. All of the different possible values of the mean differences and their associated probabilities are referred to as a **sampling distribution.** Every statistic has a sampling distribution, so we are now focusing on the sampling distribution of a mean difference. This type of distribution can be depicted as a graph or as a table, just like any frequency distribution. So it sounds as though all we have to do is to read off the probability value that corresponds with a sample difference of .8. But where do these probability values come from, and how can they be expressed in a general form that does not refer just to well-being differences on a particular 5-point scale?

This is where statistical theory comes to bear in a powerful way. Statisticians have made extensive study of so-called **theoretical distributions.** We have already encountered some of these—the normal or z distribution, and the *F* and chi^2 distributions. For each distribution, statisticians have calculated precisely the probability of any given value of the test statistic occurring and the conditions under which these probabilities are accurate. (This is the type of information often found in tables at the back of statistics textbooks.) The key step we now take is to assume that the sampling distribution of a mean difference has the same shape as a particular theoretical distribution. It turns out that in the present case we can consider either the *F* distribution or the *t* distribution as a source for the probability we seek; both will produce the same answer. Another way to describe what we have just done is to say that we have chosen a particular **test statistic,** *F* or *t,* as the statistical tool for evaluating the null hypothesis.

Before we actually find the conditional probability of the sample difference of .8, it is important to reiterate a point that slid by in the last paragraph: that statisticians have established the conditions under which test statistic probabilities are accurate. Put another way, test statistic probabilities only map accurately onto sampling distributions if certain conditions or assumptions are met. If these assumptions are seriously violated, then probabilities will be inaccurate and an inappropriate conclusion may be reached about the status of the null hypothesis. This is such an important issue that we will delve into it further in Section 2.4 and will discuss assumptions each time we explore a multivariate technique in Part 2.

We can now find the probability of obtaining a sample difference of .8 or greater, conditional upon the null hypothesis being true, first using the *F* test statistic. In order to make use of this test statistic, we have to convert the mean difference into an *F* ratio. This we did in Subsection 1.1.2 in Chapter 1, where we calculated a value of 1.23. To find the associated probability, we also need to know the degrees of freedom for the numerator (between-groups variance) and for the denominator (within-groups variance) of the *F* ratio. Table 1.1 shows these to be 1 (the number of groups minus 1) and 8 (the number of individuals in each group minus 1 summed across the two groups), respectively. The reason these are needed is that the probabilities of *F* values are different according to the number of groups and cases in the analysis. (Note that to calculate the between- and within-group variances, the corresponding sums of squares are divided by the appropriate degrees of freedom terms rather than by the actual number of groups and cases. This simple adjustment of the sample variances produces more accurate estimates of those in the populations.) All of this information is now sufficient for a computer program to calculate the conditional probability or for us to consult a table of *F* values. The result in our example, usually referred to as the ***p* value,** is a probability of approximately 30%. If the null hypothesis were true and there was actually no gender difference in well-being, a value of at least .8 would be expected in 30% of samples.

We noted earlier that we could also find the conditional probability using the ***t* test** statistic. This is because t is a special case of F and can be used when there are two groups in an analysis. So, while the t test can be used only to evaluate a difference between two groups, the F test can be used to evaluate differences between any number of groups. In the present situation, where we have two separate or independent groups, the value of t is simply the square root of F showing further their close relationship. The t value is therefore 1.11 and has 8 degrees of freedom, the same as that for the within-groups component of the F ratio. Again, consulting a computer program or a table of t values results in a p value of approximately 30%. Before leaving this subsection, two further points are worth noting briefly. First, the F or t test of a mean difference is equivalent to a test of eta. So we are simultaneously testing the significance of a mean difference and of the correlation between the independent and dependent variables. Second, if we had decided to adopt a directional alternate hypothesis, the conditional probability would just be the nondirectional p value divided by 2: 15% in this example.

Deciding on the Status of the Null Hypothesis

Now that we know the conditional probability of the sample difference, we need to make the final move of deciding whether to accept or reject the null hypothesis. Does the test suggest that the sample difference is due to chance or not? What we need is a probability *threshold* to guide the accept/reject decision. Conventionally, this threshold is set at a p value of 5% or less. Only if the p value generated by the test statistic is at or below this threshold is the null hypothesis rejected. Clearly, in the present case the p value of 30% is well above the conventional threshold, so the null hypothesis should be accepted. Essentially, we are concluding that the sample difference is so probable under the null hypothesis that there is insufficient evidence for anything other than a chance interpretation. A difference this small between groups of this size is simply not trustworthy.

The p value threshold is called the **significance level** or **alpha (α)** and may be expressed as a percentage or it may be divided by 100 and expressed as a value between 0 and 1. Statements such as "Hypotheses were tested with an alpha of .05 (i.e., 5%)" and "This result was significant at $p < .05$" are a commonplace of research reports. Another useful interpretation of the significance level is in terms of making a wrong decision about the status of the null hypothesis. One understandable concern of the analyst is the risk of rejecting the null hypothesis when in fact it is true, in the present case concluding that there is a trustworthy difference when there is not. This particular pitfall is known as a **Type I error,** and alpha quantifies the risk of making such an error. So in adopting an alpha of .05 or 5%, we are setting the level of risk that we are prepared to take of committing a Type I error.

To this point we have described the decision about the status of the null hypothesis in terms of comparing a calculated *p* value with a threshold *p* value, namely alpha. It is also possible to frame the decision in terms of threshold or **critical values of a test statistic.** This starts from the alpha value, usually .05 or 5%, and asks what value of the test statistic would be associated with this *p* value. So, for the *F* test conducted earlier we can ask what value of *F* is associated with a *p* value of .05 or less when the degrees of freedom are 1 and 8. Consulting a table of *F* values shows that the critical value of *F* is 5.32. Now the decision on the fate of the null hypothesis can be framed as a comparison between the calculated and critical *F* values. The null hypothesis can be rejected if the calculated value is equal to or greater than the critical value. Since the calculated *F* value of 1.23 is far less than the critical value of 5.32, we arrive at the same conclusion as before, that the null hypothesis should be accepted. The procedures of comparing calculated *p* values with alpha and calculated test statistic values with critical ones are equivalent and so lead to the same conclusion. As computerized analyses have become more widespread, the *p* value comparison has become more widely used. This is to the advantage of the reader since an exact *p* value is inherently more informative than a critical test statistic value, which is tied to one predetermined *p* value.

Looking back, we began this whole process by suggesting that one reason for doubting the trustworthiness of a statistical result is that it may be due to chance. This chance interpretation was then translated into a null hypothesis. The outcome of the test of this null hypothesis hinged critically on alpha—the risk of making a Type I error. In other words, we have given pride of place to the danger of rejecting the null hypothesis when it is in fact true: in the present example, concluding that there is a gender difference in well-being when there is no such difference. But what about the danger of concluding that there is no difference when in fact there is? This is known as **Type II error:** more formally, accepting the null hypothesis when it is in fact false. The outcome of our test suggested that we should accept the null hypothesis, but how much danger are we in of making a Type II error?

The degree of Type II error in an analysis is referred to as **beta** and, unlike alpha, which is simply chosen, it can be calculated. In practice it is more usual to calculate and report the value of beta subtracted from 1, or from 100 if percentages are used. This value is known as the **statistical power** of an analysis. So, for example, a beta of .20 or 20% becomes a power of .80 or 80%. The power of an analysis refers to its capability to avoid Type II error or, less enigmatically, to detect a difference or relationship that is actually present in the population. Calculating power values requires three types of information: the alpha level that has been chosen, the sample size, and an estimate of the magnitude of the difference or relationship in the population. This last piece of information is known as the **effect size,** which for our purposes we will simply refer to as "small," "medium," or "large." With this information, we can consult power tables in statistics books or run computerized power analyses to derive a power value for a particular analysis. For our analysis of 5 women and 5 men, with an alpha of .05, and an expected small effect size (suggested by the literature), we find that the power is approximately 5%. In other words, this analysis carries an approximate 95% risk of concluding that there is no difference when in fact there is. Now that is an untrustworthy result!

We can begin to understand this powerless state of affairs if we reflect on how alpha, sample size, and effect size influence the level of power achieved. The first point to note is that the harder we try to reduce Type I error, the more we increase Type II error, and vice versa. This should make sense in general terms because it says that the more we try to avoid the mistake of claiming a difference when it is nonexistent, the more likely we are to miss a difference that is actually present. Put more succinctly, reducing the value of alpha increases the value of beta and therefore decreases power. Since the value of alpha is chosen to be 5% by convention, we are free to alter it and may be tempted to increase it in order to increase power. However, doubling alpha to 10% and changing nothing else would still increase power only by about the same amount in this instance, so we would still be contemplating an alarming Type II error rate of about 90%. Moreover, the conventional choice of an alpha of 5% has been supported by considerable experience suggesting that it represents a good way of balancing Type I and II error in many situations. There are times when choosing an alpha of other than 5% may be desirable and defensible, but it is generally not advisable to view manipulating alpha simply as a means of enhancing power.

The second point to note in the determination of power is that increasing sample size reduces both Type I and II error simultaneously. Again this should make sense in general because, the more we know about what is going on in a population, the less likely we are to reach false conclusions about it. This reinforces the earlier observation in Section 2.1 that, all other things being equal, it is desirable to use sample sizes that are as large as possible. One of the fundamental reasons for the inadequate power in our example is the very small number of cases (5) in each group. This problem is compounded by the third noteworthy point about power, namely, that the smaller the effect size, the more power that is needed to detect it. In this instance we are trying to detect a small effect size using a very small sample—the worst of all worlds in power terms. Even sophisticated researchers are surprised to find how many cases are needed to achieve adequate power when a small effect size is expected. Adequacy in power terms is conventionally defined as a minimum value of .8 or 80%; that is, a beta or Type II error rate of .2 or 20%. To achieve this level of power with an alpha of .05 for a small effect size in our example would require about 400 women and 400 men! If we were seeking a large effect under the same circumstances, we would need only 26 individuals in each group, but this is still well above the sample sizes we actually used.

Even though we have been only skimming the surface, it should be clear by now that evaluating the role of chance using null hypothesis testing is a complex affair. Formulating hypotheses about parameters, calculating conditional probabilities, and choosing acceptable risk levels for drawing false conclusions about the null hypothesis all unfold to reveal hosts of subissues. As a result, it is often not possible to give a straightforward answer to the question of whether a given result is due to chance. In our example we first concluded that the gender difference in well-being of .8 was likely to be due to chance since the test led to our accepting the null hypothesis. However, we subsequently cast doubt on this analysis because the power level indicated that the risk of falsely accepting the null hypothesis was approximately 95%. The safest course under these circumstances is to treat the analysis and the result as untrustworthy mainly, as we noted, because of the doomed attempt to detect a small effect with a tiny sample. This was an extreme, contrived example, but it is worth noting in conclusion that, according to a number of methodologists, the research literature in the behavioral sciences and probably beyond is riddled with Type II error mainly owing to inadequate sample sizes. This is discouraging in some ways but also has two more positive implications. The first is that many hypotheses that seem to have failed may in fact be correct. The second is that many data analyses could be greatly enhanced by the simple expedient of investing in more cases.

Testing Hypotheses in Simple Regression and Contingency Table Analyses

Now that we have reviewed the fundamentals of null hypothesis testing using the mean gender difference in well-being as an example, we can quickly apply the procedure to the other analyses we conducted in Chapter 1. In Subsection 1.1.3 we used a simple regression to analyze the effect of positive affect on well-being and found a slope of .94. Following the same pattern as above, we specify the null hypothesis that the slope in the population is actually zero, the alternate hypothesis that it is nonzero, and an alpha of .05. The sampling distribution of a slope can be approximated with either a t or an F distribution, so either test statistic can be adopted. Using either generates a conditional probability for the calculated slope of less than or equal to .002 (.2%). Since this is far smaller than the threshold of .05 (5%), we can reject the null hypothesis. The slope value of .94 found in the sample is too large to be consistent with an actual value of zero and, in that limited sense, unlikely to be due to chance. In a simple regression this procedure simultaneously tests the null hypothesis not only for the slope, but also for r and r^2. So the F value we calculated earlier as 18.8 with 1 and 8 degrees of freedom is that used to generate the p value of .002. Whether we frame the null hypothesis in terms of the slope, r or r^2, the analysis and the outcome are identical and therefore equivalent.

In Section 1.2 of Chapter 1 we conducted a contingency table analysis and found that women were almost 6 times as likely as men to report that they were happy. Following the NHT procedure, we specify the null hypothesis that this odds ratio is 1 in the population (an odds ratio of 1 indicates the complete absence of a relationship), the alternate hypothesis that it is not 1, and an alpha of .05. The test statistic in this situation is chi^2, and the test performed is the chi^2 test of independence, that is, a test of *no* relationship. As we found earlier, the chi^2 value for this analysis was 3.33. The degrees of freedom here are derived not from the number of cases but from the number of contingency table rows minus 1 multiplied by the number of columns minus 1. Since there are 2 rows and 2 columns, the degrees of freedom are 1. A chi^2 value of 3.33 with 1 degree of freedom has a conditional p value of approximately .07. Accordingly, the null hypothesis has to be accepted despite the closeness of the .05 threshold. Even though the odds ratio of about 6 sounds impressively large, it is small enough to be interpreted as due to chance. Note that this test of the odds ratio is equivalent to testing the null hypothesis that the phi coefficient is zero. So again our earlier finding that gender and well-being were correlated approximately .41 has to be treated as due to chance according to the NHT procedure. That said, remember that the issue of power may provide a caveat to this conclusion. Without our going into detail, the power of this analysis was certainly less than the desirable minimum of .8, but far better than in our analysis of a mean difference.

This completes our rapid but hopefully sufficient introduction to null hypothesis testing. We will return to the topic at the end of Chapter 3 when we review the strengths and limitations of the foundations of multivariate data analysis. Meanwhile, in the next section we turn to a different way of evaluating the role of chance in determining the trustworthiness of statistical results—a procedure called estimation.

2.3.2 Evaluating the Role of Chance With Estimation

Null hypothesis testing has been criticized on a number of grounds. One of the fundamental concerns is the reduction of statistical decision making to a simple accept or reject choice, especially given the unlikely nature of what is being accepted or rejected, that is, the total absence of an effect. To meet this sort of concern, it would be helpful to somehow make use of the statistical tools used in the actual testing phase of the NHT procedure, but to jettison the hypothesis-testing framework. This way of thinking leads us to another form of statistical inference called **estimation.** In essence, this involves estimating the value of a parameter using sample statistics, without any preconceptions about what that value might be. So, returning to our gender mean difference in well-being of .8, we could use this sample statistic to estimate the mean difference in the population. Similarly, we could estimate the population slope for the effect of positive affect on well-being or the population odds ratio for the gender difference in happiness. Any statistic can be used as a basis for estimating its corresponding parameter.

If all we are seeking is a single value for the parameter of interest, we are engaging in **point estimation.** In the case of the mean difference, the slope, and the odds ratio, there is no more work to be done because the sample statistic simply is the best estimate of the parameter. Our best point estimate of the population gender mean difference is .8—the value of the sample mean difference. Similarly, the sample slope of .94 and the sample odds ratio of 5.97 are the best point estimates of their corresponding parameters. Usually, though, a more cautious approach to estimation is used, called **interval estimation.** This still takes the point estimate as the starting point but then calculates a range of values within which the parameter value is likely to lie with a specified level of risk, now called confidence. The range of values is referred to as the **confidence interval,** and the degree of risk is known as the **confidence level.** For example, in a moment we will calculate that, while the best point estimate for the gender mean difference in well-being is .8, we can have 95% confidence that the parameter value lies between –.86 and +2.46. But where do these figures come from?

Earlier we noted that estimation makes use of the same statistical tools found in the NHT procedure. This procedure hinges on the calculation of a test statistic such as *F, t,* or chi^2. Although it is not always apparent from calculation formulae for test statistics, they always have three components: the sample statistic value, the hypothesized parameter value, and something called the **standard error** of the sample statistic. The sample statistic value represents the effect of interest, and the parameter is set to the value specified in the null hypothesis, usually zero. The standard error indicates how variable the sample statistic is when repeated samples are taken from the parent population. It is in fact the standard deviation of the sampling distribution of the sample statistic and can be calculated from the sample standard deviations.

In NHT, appropriate values for the sample statistic, parameter value, and standard error are combined to derive a value for the test statistic. Interval estimation rearranges these elements and combines values for the sample statistic, standard error, and test statistic to produce an interval estimate for the parameter. The sample statistic and standard error are calculated from the sample data as in NHT. The value for the test statistic is chosen to reflect the desired level of confidence and the degrees of freedom. This confidence level can be thought of as the opposite of risk, as expressed in an alpha level. Since the conventional alpha level is 5%, the corresponding conventional confidence level is 100 – 5 = 95%.

Putting all this together, we can calculate 95% confidence intervals for our mean difference, slope, and odds ratio statistics. As noted above, the mean difference of .8 has an associated confidence interval such that we can be 95% confident that the population mean difference lies between −.86 and +2.46. This wide range (for a 5-point scale) reflects the small sample size and the variability in the sample. Both combine to make it hard to come up with a precise estimate of the parameter value. It is also worth noting that this confidence interval includes the value of zero. This is consistent with our NHT conclusion that the parameter value could be zero, that is, we accepted the null hypothesis on the basis of an alpha of 5%. Turning to the slope representing the impact of positive affect on well-being, we see that the 95% confidence interval turns out to range from .44 to 1.43, centering on the point estimate of .94. Note that this time the interval does not include zero and is again consistent with our NHT decision, on this occasion to reject the null hypothesis. Finally, the 95% confidence interval for our odds ratio lies between .81 and 44.35, with a point estimate of about 6. This enormous range again reflects the imprecision generated by this inadequate data set; and the fact that the interval includes 1 is consistent with accepting the null hypothesis of no relationship.

This very brief introduction to estimation should be sufficient at least to suggest the merits of statistically evaluating chance without hypotheses. Interval estimation provides all of the information found in NHT but offers a more detailed chance assessment without a forced binary decision about the status of an intrinsically unlikely hypothesis. Despite attempts to replace NHT with interval estimation, the latter still holds sway as the preferred way to evaluate chance, at least in psychology. In Part 2 we will make use of both approaches as we examine multivariate techniques. But in recognition of the continuing dominance of NHT and the need for readers to make sense of its use in the mainstream literature, interval estimation will be far less evident in our discussions.

2.4 STATISTICAL ASSUMPTIONS

Like any tools, statistical methods can be misused. As we have seen, their application to samples of inadequate size, or to data generated with an inappropriate measurement scale, may invalidate their use. But even if the sampling and measurement strategies are appropriate, a further set of issues arises when we engage in null hypothesis testing or estimation to evaluate the role of chance. These procedures depend crucially on the generation of accurate probabilities—*p* values or confidence levels—through the calculation of test statistics. The accuracy of these probabilities may be distorted if certain conditions or assumptions are not adequately met. In this final section on trustworthiness we introduce three assumptions that arise initially in even simple inferential analyses and then subsequently in varying forms in more complex analyses. The assumptions revolve around the independence of cases, the normality of frequency distributions, and the equality of variances across distributions. We will discuss each of these in general terms and with specific reference to our earlier analyses of the difference between means and of correlation and regression coefficients in particular.

2.4.1 Independence of Cases

All of the inferential analyses we have encountered so far, using *F, t,* and chi^2 statistics, require that each individual's data are independent of every other individual's. In our examples this means that each individual's well-being score must be unaffected by the score of anyone else in the sample. How might some linkage or dependency among scores come about? In the extreme case an individual could appear twice in the data set, perhaps through a misguided desire to help the investigator increase statistical power without having to recruit more participants. Clearly, this person's two scores would not be independent of each other, even if they were not identical. But more subtle processes are possible and more likely. For example, some of the participants in these well-being samples may be partners or siblings whose well-being experiences are in some way intertwined. Dependency may also occur because of changes in data collection procedures over time. Perhaps, as the researcher becomes more experienced and empathetic in instructing participants on how to complete the well-being measure, individuals tested later in the sample may admit to lower levels of well-being compared with those tested earlier. If we broaden our focus to include experimental studies, then more potential for dependency of scores emerges. Imagine an experimental study in which the objective is to test the effect of a potentially mood-enhancing stimulus, such as a funny movie, on groups of participants. Obviously, the fact that this experiment is constructed as a group experience means that individual responses are unlikely to be independent of each other.

Given that there are a number of ways in which dependencies across individuals can occur, why does it matter? It matters because there is good evidence to suggest that even a small degree of dependency can distort *p* values in a predictable and worrying way. Specifically, the consequence is in effect to inflate alpha. Our analysis may appear to be protected by the conventional 5% risk of making a Type I error, but the presence of dependent data can multiply this level several times over. Stevens (2002) provides a good, accessible discussion of this issue and concludes that failing to meet the independence assumption can be very serious. This conclusion means that knowing how to detect dependent data and how to deal with them are important issues for a conscientious data analyst. Accordingly, we will look more closely at them in later chapters, particularly in Chapter 4 on multiple regression.

2.4.2 Normality of Distributions

A further assumption required by our earlier ANOVA and regression analyses is that the dependent variable be normally distributed in the populations. Note that this requirement concerns what is assumed to be occurring in populations, not in samples. Although the frequency distribution of well-being in the sample may give some indication of the corresponding population distribution, it is the latter that is of interest. Note further the reference to population in the plural. In the case of ANOVA the assumption requires that the distribution of well-being is normal both in the population of women and in that of men. For regression, the distribution of well-being is assumed to be normal in every population defined by each value of the independent variable. So the population distributions required to be normal are those from which subgroups of cases are sampled, one sample for each level or value of the independent variable such as gender or positive affect. Unlike the independence assumption, the normality assumption does not apply to chi^2 analyses. Chi^2 used in this way is what is called a **distribution-free statistic,** meaning that its application requires no assumptions of any sort about how variables are distributed in the populations sampled.

In later chapters we will frequently encounter the normality assumption. It is encouraging to discover at the outset then that it is a so-called **robust assumption.** This means that even when population distributions deviate quite markedly from normality, *p* values suffer little distortion and the risk of a Type I error is close to what it appears to be. There are two caveats to this reassurance, though.

First, the robustness deteriorates with very small sample sizes and rapidly improves as the sample size increases. So we arrive again at the wisdom of acquiring good sample sizes by another route. Second, the fact that the mean and statistics derived from it are most informative when they are describing a normal distribution suggests that there are other reasons for worrying about nonnormal distributions than just erroneous p values. In practice, as we will see in later chapters, it is possible to mathematically transform distributions to directly reduce nonnormality and to indirectly counter other assumption violations.

2.4.3 Equality of Variances

The final assumption required by many inferential analyses relates to the variances of the population distributions that have been sampled. These distributions are required not only to be normal, but also to have equal variances. In our ANOVA example this means that the variance of well-being in the women's population should be equal to the variance in the men's population. For the regression analysis, the variance in well-being scores should be equal for all population subgroups sampled for each value of the positive affect variable. Depending on the analysis, this assumption may be labeled as **equality of variance, homogeneity of variance,** or **homoscedasticity.** One way to think about a rationale for this assumption is to appreciate that calculating F and t statistics involves pooling or averaging variances of the dependent variable across categories or values of the independent variable. For example, our earlier F test of the mean well-being difference for women and men required us to pool the within-group differences. If the pooled variance hides a marked disparity in the group population variances, the value of F and therefore p may be distorted.

The equality of variance assumption is also robust, but only when the group sizes in the sample are equal or approximately equal. Stevens (2002, p. 268) recommends that the largest group should be no more than 1.5 times as big as the smallest group as a way of defining the limits of "approximately." If there is evidence that variances do differ, and groups are not at least approximately equal, it is possible that Type I and II error rates may be affected. Accordingly, in later chapters we will explore ways to detect inequality of variances and to rectify the situation. Note again that, since chi^2 is a distribution-free statistic, the assumption on equality of variance is not relevant to its application.

This completes our brief introduction to the role of statistical assumptions in evaluating trustworthiness. The focus has been on three assumptions that are required by many inferential analyses if the test statistic and *p* values are to be accurate and credible. One other fundamental assumption also deserves mention in this introduction, even though it is not strictly required for accurate inference as such. This is the assumption of **linearity,** which earlier we mentioned in passing in the context of regression analyses. Linearity is the assumption that the most effective way to summarize the profile of mean differences in a dependent variable across levels of an independent variable is with a straight line. If this assumption is tenable, the payoffs are considerable. For example, assuming linearity in the relationship between positive affect and well-being enabled us to calculate linear regression coefficients, notably the slope of .94. This one number, simply calculated, captures the impact of positive affect on well-being regardless of the values of the variables. If the relationship is linear, the slope does not change across the range of the variables' values. So the linear slope is a highly informative statistic in that it neatly encapsulates a relationship that applies across all scores in the sample.

On the other hand, linearity should not be assumed just because it leads to simple outcomes. Many relationships in the behavioral and social sciences are likely to be better summarized with some nonlinear profile. Good observation and good theory are needed to decide which these might be and what form they might take. Even in the absence of these, any data set should be examined to see if the assumption of linearity is justified. In later chapters we will explore methods for checking linearity and for dealing with situations where nonlinearity is suspected.

2.5 FURTHER READING

A good introduction to sampling strategies is provided by Kalton (1983), while accessible discussions of measurement issues are available in Section I of Kline (1993) and more extensively in Nunnally and Bernstein (1994). The mechanics of null hypothesis testing and estimation are covered in more or less detail in all of the references cited at the end of Chapter 1. Cohen (1988, 1992) and Kraemer and Thiemann (1987) discuss in an accessible way the particular topic of statistical power. Good discussions of the issues involved in evaluating statistical assumptions can be found in Tabachnick and Fidell (2001, Chapter 4), and Hair, Anderson, Tatham, and Black (1998, Chapter 2).

THREE

ACCOUNTING FOR DIFFERENCES IN A COMPLEX WORLD

This chapter provides a ladder that leads from the foundations laid in the first two chapters to the higher levels of data analysis where we find multivariate techniques. The ideas and techniques we have discussed so far enable analysts to find patterns in one or two sets of differences, that is, to engage in univariate and bivariate analysis. However, it is obvious that the world is not neatly packaged into isolated pairs of variables. So if attempts are made to account for differences only in a two-by-two fashion, it is inevitable that the resulting pictures will be at best limited or at worst distorted. Multivariate techniques are tools that help the analyst reduce these limitations and distortions by capturing more complex portions of the world inside one analytic frame.

In Section 3.1 we examine the fundamental limitations of bivariate analysis when it is applied to data representing three or more variables. These limitations stem partly from the increasing number of variables as such, but also from the more complex patterns of relationships that now become possible and even likely. This latter concern will lead us to introduce the distinctive relational patterns of confounding, moderating, and mediating relationships. Section 3.2 introduces the strategy that lies at the heart of multivariate data analysis, and discusses in general terms how this strategy combats the limitations of bivariate analysis identified in Section 3.1. The strategy involves combining variables into a composite by "weighting" each variable and adding it into the composite. The notions of composites and weights are so fundamental that we will examine their meaning and interpretation in some detail, but always stressing conceptual rather than statistical issues. To ensure that these central ideas are firmly grounded, we begin the section with a review of simple regression; extend this into multiple regression, which accommodates more than one independent variable; and finally show how the multiple regression approach can be generalized into a broad MDA strategy. In Section 3.3 we step back from the details and try to gain a more critical perspective on what multivariate data analysis has to offer. Its strengths are considerable, but they can easily be exaggerated or at least misinterpreted. This happens partly because of the language employed, when statistical terms such as "explanation" and "prediction" appear to promise a lot but actually have much more constrained meanings than they do in general research discourse. With these final reflections in mind, we should then be ready to approach the techniques themselves in Part 2 in an informed and balanced way.

3.1 LIMITATIONS OF BIVARIATE ANALYSIS

In Chapter 1 we used simple regression to explore the possibility that differences in positive affect might account for differences in well-being. Since we also had data on life satisfaction, we could have repeated the analysis to see whether differences in satisfaction also account for differences in well-being. What are the limitations of addressing these two relationships in separate bivariate analyses? They can be organized into three categories: descriptive, inferential, and relational limitations.

From a descriptive perspective, we would have no indication of how positive affect and satisfaction are *jointly* related to well-being. For example, we would have no ready answer to the question of how much variance in well-being is explained *in total* by these two attributes. Under certain circumstances an answer to this question can be derived from bivariate analyses but, as we will see, these circumstances are likely to be rare. More powerful answers to this type of descriptive question are to be found in multivariate analysis.

Turning to statistical inference, in Chapter 2 we found that the outcome of a null hypothesis test depends critically on alpha, the threshold probability that must be met if the null hypothesis is to be rejected. The value of alpha is our protection against inappropriately rejecting the null hypothesis (Type I error) more often than we would wish—usually 5% of the time. One feature of this procedure that we passed over at the time is that the calculations are predicated on a "one-off" sampling logic. If we keep returning to the data set to test more and more hypotheses that are linked by common variables, the actual value of alpha inflates invisibly. So, while we may think that a protective alpha of .05 is in place, its real value may be much higher. Put another way, the more hypotheses we test about the same variables, the more likely we are to believe erroneously that we have found an effect. A wide range of techniques have been developed to counter this inflationary consequence of multiple tests, and we will explore some of them in Chapter 6. These are valuable, but it would be helpful to also have at our disposal another approach that enabled us to test a *set* of hypotheses all at once. We will find that multivariate analysis provides exactly this facility so that, for example, we could evaluate whether positive affect and satisfaction have any statistically significant relationship with well-being using just one multivariate test.

Although the limitations of bivariate analysis with respect to indexing the total effect of a set of variables and to reducing the number of hypothesis tests are important, those concerning complex relationships among variables easily outweigh them. Here we return to the central idea that the world is not composed of isolated pairs of variables. It is unlikely, for example, that any effects of positive affect and satisfaction on well-being are independent of each other, or indeed of other variables so far unmentioned. If this is true, we need to think systematically about how three or more variables might be related in principle and then about how such patterns might be statistically analyzed. Clearly, by definition, bivariate analysis will not be up to the task, and so we reveal another motive for turning to multivariate analysis.

Three types of multivariable relational patterns are commonly considered by analysts: the confounding, moderating, and mediating patterns. We can illustrate all of these with our two independent variables and one dependent variable, but we need to appreciate that these are the simplest, three variable versions. As more variables are added into an analysis, the potential for ever more complex relationships grows at an alarming rate. Also, this basic list of patterns is not exhaustive, but it covers many analytic situations and provides building blocks for other possibilities.

In the **confounding** pattern the individual effects of two independent variables on a dependent variable are distorted because the independent variables are themselves related. So, for example, if positive affect and satisfaction are systematically related in some way, then it may prove difficult to untangle the effect of each on well-being: They are potentially confounded with each other. This means that a search for confounds should focus on variables that might be systematically related to an independent variable *and* to a dependent variable: a *triangle* of relationships. It is important to appreciate, though, that even when such a triangle is suspected or demonstrated, distortion due to confounding will not necessarily occur. The relationship triangle is a necessary but not sufficient condition for confounding. It is also important to note that distortions due to confounding can manifest in a variety of ways. Depending on the strength and direction of the relationships in the triangle, effects may appear, disappear, increase or decrease in magnitude, or change their sign. As a consequence, it is desirable not only to consider and measure potential confounding variables, but also to conduct analyses that reveal where confounding is actually occurring and to exercise appropriate control over it. Multivariate analysis provides the means to achieve this, using what is known as statistical control. Later we will use this type of control to see if positive affect and satisfaction are confounded and to obtain estimates of their unconfounded effects on well-being.

The **moderating** pattern is not a form of potential distortion as such, but it opens up the way for more sophisticated theorizing even with only three variables. In essence it suggests that the relationship between an independent variable and a dependent variable differs according to the *level* of a third variable. It is thus suggested that the third variable *moderates* the relationship. In our example we might suggest that the impact of positive affect on well-being will be stronger for individuals who enjoy a higher level of satisfaction. The key word here is "stronger" as this specifies how the effect changes at higher levels of the moderating variable. This type of pattern in which a bivariate relationship is thought to differ according to the level of a third variable is also known as an **interaction effect.** In the example positive affect and satisfaction are proposed to have an interactive or joint effect on well-being, which is additional to whatever individual effects each may have. Here we have a so-called *two-way interaction effect* since two independent variables are implicated. As more independent variables are added to the picture, more two-way and higher-order interaction effects become possibilities. Finally, note that the analyst is free to specify any form of moderating relationship as long as there is theoretical or empirical justification. The particular form suggested here, whereby the effect of positive affect *increases* at *higher* levels of satisfaction, has just been drawn from the air, however much it might appeal to common sense.

The **mediating** pattern can be thought of as a causal chain and as another way in which theorizing might be made more sophisticated. In this pattern an independent variable is seen as having an effect on a dependent variable *through* another independent variable. Put another way, if we imagine a causal chain linking three variables, the middle variable is said to mediate the effect of the first variable on the third. We might theorize that the effect of satisfaction on well-being is mediated by positive affect. In this case, satisfaction produces positive affect that in turn enhances well-being. It is possible in principle to distinguish between total and partial mediation. Applying the total mediation scenario to our example would suggest that the *only* way in which satisfaction influences well-being is through positive affect. Partial mediation would suggest that positive affect is only one pathway by which satisfaction might have an impact on well-being. Choosing between total and partial mediation effects is again the prerogative of the analyst and an important issue as it will guide expectations about the patterns that should emerge in multivariate analyses.

Later in the following section we will discuss in general terms how confounding, moderating, and mediating relationships can be viewed from a multivariate analysis perspective. For now, it is important mainly to appreciate the form of each type of relationship and the distinctions among them. It is also important to appreciate that the types of relationship are not mutually exclusive. All three may occur in the same analysis and require appropriate analytic strategies. Whatever else is clear at this point, it should be evident that bivariate strategies in and of themselves cannot begin to deal with the complexities we have introduced. They do, however, provide the building blocks from which multivariate strategies can be constructed.

3.2 THE MULTIVARIATE STRATEGY

We are now finally ready to approach the multivariate question: How do we set about accounting for differences when the analysis contains three or more variables? As we saw in the last section, ideally the answer to this question should enable the analyst to treat variables in sets or subsets, minimize the number of statistical tests, and provide ways of capturing confounding, moderating, and mediating patterns of relationships. Our discussion of the answer will begin in Subsection 3.2.1 with a review and some further elaboration of the technique of simple, or bivariate, regression analysis that we introduced in Chapter 1. Then, in Subsection 3.2.2 we will see how this can be easily extended into **multiple regression,** when there are two or more independent variables. After exploring in general terms how multiple regression can be used to analyze confounding, moderating, and mediating relationships, we will discuss in Subsection 3.2.3 how the regression approach can be seen as just one example of the generalized strategy that lies at the heart of multivariate analysis. So simple regression and multiple regression are used in this section simply as a vehicle to introduce the key concepts of multivariate analysis. A more detailed treatment of multiple regression can be found in Chapter 4.

3.2.1 A Review of Regression Building Blocks

Imagine that 5 individuals, identified as A–E, provide scores on measures of positive affect, satisfaction, and well-being. This time each score is on an interval scale with a possible range of 1 to 10. The imaginary scores appear in Table 3.1.

Table 3.1 Three Sets of Scores for 5 Individuals (A–E)

Individual	*Positive Affect*	*Satisfaction*	*Well-Being*
A	3	3	3
B	5	6	5
C	6	5	4
D	7	4	5
E	9	7	8

To what extent are differences in well-being due to positive affect in these data? As discussed in Chapter 1, the relationship between two interval-level variables can be summarized with a regression equation, which has two unknowns: the slope and the Y intercept. Calculating these for the present data gives a slope of .75 and a Y intercept of .5. These numbers can be used to address the research question from the perspective of individual differences and of group differences. But before we look at these, it is helpful to begin by focusing on just one individual such as person E. The regression equation can be used to predict Person E's well-being score on the basis of E's positive affect score of 9, by multiplying the latter score by the slope and adding the Y intercept, as follows:

E's predicted well-being score = slope times E's positive affect score
+ Y intercept
= .75(9) + .5
= 7.25

How much in error is this prediction for person E? This can be calculated by simply subtracting the predicted from the actual well-being score of 8: $8 - 7.25 = .75$, an error value known as the residual. So, in effect, the regression equation can be used to generate two new scores for person E or for any other individual in the sample: a predicted score on the dependent variable and a residual. The first score of 7.25 represents that part of the dependent variable that can be predicted from the independent variable, and the second residual score of .75 represents the remaining unpredictable part. Moreover, obviously these two new scores add up to the dependent variable score:

E's dependent variable score = E's predicted score
+ E's residual score
8 = 7.25 + .75

Since each person has a predicted and a residual score, we can summarize across the 5 cases and derive statistics that capture individual differences in the usual way. All of the relevant statistics appear in the ANOVA summary table in Table 3.2. The sum of squares for the predicted scores is otherwise known as the regression sum of squares that we encountered in Chapter 1, and dividing it by the number of independent variables produces the regression variance (mean square). In the present example the regression sum of squares and variance are both 11.25. How can this be turned into a more interpretable statistic that indicates how much of the individual differences in well-being can be accounted for by differences in positive affect?

Table 3.2 ANOVA Summary Table Showing the Relationship Between Positive Affect and Well-Being

Source of Differences	*Sum of Squares*	*Degrees of Freedom*	*Variance*
Regression	11.25	1	11.25
Residual	2.75	3	0.92
Total	14.00	4	

The regression sum of squares of 11.25 indicates how much of the individual differences in well-being can be predicted from differences in positive affect. The *total* amount of individual differences in well-being regardless of any other variable is indicated by the sum of squares for well-being, which is 14. These two figures turn out to be directly comparable because of the Y intercept, about which we have had little good to say up to this point. The Y intercept can be seen as an adjustment to the regression calculations, which ensures that the mean of the predicted dependent variable is equal to the mean of the dependent variable itself. It then follows that our two sums of squares numbers are comparable because they are based on deviations around the same mean. As a result, we can simply divide the regression sum of squares by the total sum of squares and arrive at r^2: the proportion of variance in the dependent variable accounted for by the independent variable. For the present data r^2 is 11.25/14 = .804, which says that positive affect accounts for 80.4% of variance in well-being.

Turning now to the five residual scores, we see that these can also be turned into a sum of squares and variance. Following from the logic of r^2, the residual sum of squares divided by the total sum of squares will indicate the variance in well-being that is *not* accounted for by positive affect. This number is 2.75/14 = .196, which says that positive affect fails to account for 19.6% of the variance in well-being. As we would expect, explained variance and unexplained variance add up to 1 or 100%. It is also possible to view unexplained variance in terms of a standard deviation. If we take the square root of the residual variance, we produce the standard error of estimate, which in the present case is .96. This is directly comparable with the standard deviation of the dependent variable, which is 1.87. One way to make this comparison is in terms of a proportional reduction in error. If we subtract the standard error of estimate from the standard deviation of the dependent variable and divide it by the latter, we can see proportionately how much error is reduced by taking account of the independent variable as opposed to just using the mean of the dependent variable. The calculation here is (1.87 –.96)/1.87 = .487. This means that the amount of error made in predicting individual well-being scores is reduced by 48.7% if we take account of positive affect scores.

We have used the regression equation to generate predicted happiness scores and residuals for each individual and then derived summary statistics such as r^2 and the standard error of estimate to quantify how far *individual* differences in positive affect account for *individual* differences in well-being.

Now we turn to the question of how far *group* differences in positive affect account for *group* differences in well-being. This is simply answered by reference to the slope statistic in the regression equation. The value of .75 says that groups who differ by 1 on the positive affect measure differ by .75 on the well-being measure, on average. Note that because we are now taking a group perspective, we are concerned with what's happening "on average."

This completes the review of key statistics in simple regression. We have taken time to revisit and elaborate on them because all of them carry over into multiple regression. So, having a good grasp of them at this point will smooth the transition we are about to make from bivariate to multivariate analysis.

3.2.2 The Composite Variable in Regression

In Subsection 3.2.1 we divided up the relationship between two variables, well-being and positive affect, into additive components. We saw that:

well-being score = predicted well-being score
+ residual

and decomposed the right-hand side of the equation into:

predicted well-being score = slope(positive affect score)
+ Y intercept

It is this regression equation that provides the vehicle for dealing with more than one independent variable. All that is required is to extend the equation by literally adding in a new term for each additional independent variable. So, for the data in Table 2.1 the equation would be:

predicted well-being score = slope(positive affect)
+ slope(satisfaction)
+ Y intercept

To explore the details of this new multiple regression equation, we will focus again on person E in Table 3.1. The regression equation for person E is:

E's predicted well-being score = (.5)(9) + (.5)(7) + (–.5)

The scores for positive affect (9) and satisfaction (7) are taken from the last row of Table 3.1. The values for the slopes (which happen to be the same) and Y intercept have been calculated using the SPSS computer package. (As we move into multivariate analysis proper we will no longer refer to the calculations as such as they become too complex and are best left to a computer.) The predicted well-being score for person E is 7.5. Since this person's actual well-being score is 8 (see Table 3.1), the residual score is .5. Note that the predicted score is now based on two independent variables rather than just the one we used earlier and, for this case, has reduced the residual from .75 to .5. Adding in information about person E's satisfaction has increased predictive power or, equivalently, reduced predictive error. This may be true for this person E, but how well are we now accounting for individual differences for all of the cases?

Although the regression equation has been extended, the predicted happiness scores and residuals are no different in form from those in simple regression. So we can proceed straight to summary statistics derived from these scores. In terms of explained variance, the computer calculations result in a value of .893. Positive affect and satisfaction together account for 89.3% of individual differences in well-being for these 5 cases, compared with the 80.4% we achieved with just positive affect. In multiple regression this statistic becomes the multivariate version of r^2 and is now referred to as **multiple R^2,** with an uppercase *R,* but its interpretation remains unchanged. The other individual difference statistic we reviewed earlier was the standard error of estimate: an indicator of unexplained variability. When only positive affect was in the equation, the standard error of estimate was .96. When satisfaction is added in, the computer calculations show that the standard error of estimate drops to .87—a further reduction in error.

Reflecting on the nature of R^2 takes us straight to the core concept in multivariate analysis. It can be thought of as the squared correlation between the actual and predicted well-being scores. Since the predicted scores are generated by the independent variables packaged up inside the multiple regression equation, R^2 can also be seen as the squared correlation between the dependent variable and a *composite variable* that contains all of the independent variables. This forming of a composite variable in order to analyze many variables all at once is the strategy that lies at the heart of multivariate analysis. Here we see the strategy in action in the form of a multiple regression equation, but later we will generalize the idea to reveal its analytic power and scope more fully.

The composite variable not only generates the predicted and residual scores for each individual in the analysis, but also contains information that indicates in what way *group* differences in the independent variables relate to *group* differences in the dependent variable. As in simple regression, this information is to be found in the slopes. The simple regression we conducted earlier on positive affect and well-being resulted in a slope of .75, whereas the positive affect slope in the multiple regression is .5. Something has shifted, but what exactly? In a multiple regression a slope is more properly known as a **partial slope.** Roughly speaking, it quantifies the impact of an independent variable over and above the impacts of all other independent variables in the equation. The notion of a partial slope is not intuitively obvious, and it is fundamental to multivariate analysis. So it will be beneficial to explore various ways in which a partial slope can be interpreted.

The most general way to conceptualize a partial slope is in terms of statistical control. The partial slope of .5 indicates the impact of positive affect on well-being when satisfaction is statistically controlled. Similarly, the partial slope for satisfaction (which also happens to be .5) indicates its impact on well-being when positive affect is controlled. But *what* is being controlled? The answer is the relationship between positive affect and satisfaction. The correlation between these two independent variables turns out to be .71, that is, they share just over 50% of their variance ($.71^2 = .504$). This means that their effects on well-being may be confounded, and so some sort of confound detection and control is needed. This is exactly what multiple regression offers by way of partial slopes. The shift from a slope of .75 in the simple regression to a partial slope of .5 in the multiple regression shows confound detection and control in action for the positive affect variable. The first number indicates the effect of positive affect on well-being, whereas the second indicates the same effect once the confounding relationship with satisfaction has been controlled or neutralized.

Another way to approach the partial slope is in terms of holding a potentially confounding variable constant. The strategy of controlling a variable by holding it constant is a staple of experimental design. If we wanted to study the causes of a behavior that varied systematically with the time of day, for example, then it would be wise to always conduct the study at the same time. Holding time constant in this way would ensure that it could not confound the effects of other variables of interest. Turning a variable into a constant guarantees that it cannot covary with any other variable and so cannot be a confound. The confound triangle of relationships has been broken. This strategy of literally holding something constant is obviously not always a practicable or ethical option. The partial slope can be seen as another way of achieving the same end in a nonliteral, statistical fashion. So the partial slope for positive affect indicates its impact on well-being when satisfaction is held constant. Put another way, the partial slope shows how much well-being differs across groups who differ by a score of 1 on the positive affect measure, but who do not differ on the satisfaction measure.

A third way to think about the partial slope is in terms of adjusted variables. From this perspective the partial slope shows the impact of an independent variable on the dependent variable after the independent variable has been adjusted to take account of its relationships with the other independent variables. Synonymously, analysts also talk about "correcting" for or "partialing" out other variables. So the partial slope for positive affect captures the impact of positive affect on happiness after adjusting for, correcting for, or partialing out positive affect's relationship with satisfaction. Although we will not delve into the statistical mechanics of how this is achieved, it is instructive to look a little more closely at how partialing is performed. One way to do this is to think of multiple regression as a series of simple regressions.

To find the partial slope for positive affect, we could first conduct a simple regression with positive affect as the dependent variable and satisfaction as the independent variable. This would generate two new variables: a predicted positive affect score and a residual score for each case. The residuals variable contains those differences in positive affect that are *not* predicted by differences in satisfaction. So the residuals variable can be treated as an adjusted positive affect variable, that is, adjusted to exclude any differences due to satisfaction. The positive affect variable has been "residualized" with respect to satisfaction. This residualized variable would then become the independent variable in another simple regression with well-being as the dependent variable. The slope for positive affect from this simple regression would be equal to the partial slope from the multiple regression since the positive affect slope has been adjusted for the relationship between positive affect and satisfaction. We could do the same parallel procedure for satisfaction. First there would be a simple regression, this time with satisfaction as the dependent variable and positive affect as the independent variable. Then there would be another simple regression with the residuals variable from the first regression as the independent variable (residualized satisfaction) and well-being as the dependent variable. This would produce a slope indicating the impact of satisfaction having adjusted for positive affect.

Using the regression equation to form a composite variable provides solutions to some of the limitations we discussed in Section 3.1. By generating the predicted dependent variable, we have a device for calculating statistics such as R^2 that refers to the independent variables *as a set*. The multiple regression shows that positive affect and satisfaction *together* explain 89.3% of the variance in well-being. It is important to appreciate that this R^2 figure is different from the sum of the two r^2 values that are generated by the two separate simple regressions. These simple regressions show that positive affect explains 80.4%, and satisfaction explains 71.4% of variance in well-being. However, because of the relationship between the two independent variables ($r = .71$), the relationship of each with the dependent variable is inflated in this case. The positive affect variable's predictive ability is confounded with that of satisfaction and vice versa. The statistical control provided by multiple regression not only produces partial slopes, but also produces an R^2 value that is adjusted to take account of correlations among independent variables. The only time that a multiple regression R^2 is equal to the sum of the simple regressions' r^2*s* is when the independent variables are uncorrelated, and so there is nothing to adjust for. Since multivariate techniques are most commonly adopted in order to deal with correlated independent variables, this is a rare situation.

The second limitation of conducting bivariate analyses on multivariate data that we identified in Section 3.1 was the problem of the inflation of Type I error (alpha) produced by multiple hypothesis tests. This, too, is countered by the composite variable and its products. When a multiple regression is carried out, it is usual to begin by testing the null hypothesis that R^2 is zero in the population sampled using an F test. This is equivalent to testing whether *any* of the independent variables has a statistically significant relationship with the dependent variable. Since all of the relationships are tested simultaneously, this is often referred to as an **omnibus test.** This is a very efficient approach in the sense that if the test of R^2 is *not* statistically significant, that is, the null hypothesis is accepted, there is no need to conduct further hypothesis tests on individual independent variables. In our example the F of 8.33 has an associated p value of .11. Assuming that we adopt a conventional alpha of .05, the null hypothesis is therefore accepted, and no further tests are justified. Despite the large sample value for R^2 of .893, it is still consistent with the hypothesis that the population R^2 is zero. Note in passing that this outcome is highly influenced by the tiny sample size and the consequent lack of statistical power.

If the test of R^2 had been statistically significant, we could have proceeded to test hypotheses about each of the partial slopes. This type of analysis tests the hypothesis that a partial slope is zero in the population using either a *t* or an *F* test. For example, the partial slope for positive affect is .5, its *t* value is 1.83, and its *p* value is .209. As the R^2 test already suggested, there is no significant relationship between positive affect and well-being, and a *t* test for the satisfaction slope would lead to the same type of conclusion. Hypothesis testing will play an important role in the multivariate techniques that we explore in Part 2. For now, the key point to note is that the use of a composite variable allows the analyst to test multiple hypotheses simultaneously and thereby reduce the number of tests in any one analysis. This in turn helps to combat the problem of inflating Type I error, which occurs when multiple, linked hypotheses are tested in sequence. Further, hypotheses can also be tested about the partial slopes that capture the effects of independent variables after they have been adjusted for confounding relationships among the independent variables.

The third limitation of bivariate analyses is that of handling multivariable patterns such as confounding, moderation, and mediation. By now it should be clear how multiple regression provides a tool for detecting and controlling confounding relationships. It can also be used to analyze moderating and mediating relationships, and the general ways in which this is accomplished bring to the surface two further strengths of multivariate analysis. These are the capacity to represent complex effects as single terms in a regression equation and to analyze complex relationships using series of regression equations.

In our earlier discussion of moderating relationships, we suggested the possibility that satisfaction might moderate the effect of positive affect on well-being. Put another way, this is the possibility that positive affect and satisfaction might have an *interactive* effect on well-being *over and above* their individual effects. This additional effect can be thought of as another independent variable that is literally the product of the positive affect and satisfaction variables. The new variable can be added to the multiple regression equation so that it now takes the form:

predicted well-being score = slope(positive affect) + slope(satisfaction)
+ slope(positive affect X satisfaction)
+ Y intercept

Values for the partial slopes in this multiple regression equation can be found in the usual way. Our particular interest would be in the partial slope for the interaction variable. If this partial slope was statistically significant, we would have evidence that there is an interaction effect over and above the individual effects of the two independent variables. Further analysis would then be required to find out whether the particular form of the interaction was in line with expectations. Statistical details aside, the important conceptual point here is that moderating relationships can be literally added into a composite variable and tested in a controlled fashion, just like any other independent variable. Generalizing this point further, we will see that a variable in a regression equation can be constructed to represent all sorts of effects, which gives the technique enormous scope for capturing complex relationships.

Regarding mediation relationships, it was suggested earlier that the effect of satisfaction on well-being may be mediated by positive affect. This envisages a causal chain in which satisfaction leads to increased positive affect, which in turn enhances well-being. This pattern can be evaluated with a *series* of regressions that decompose the chain of relationships. Such an approach is called **hierarchical or sequential regression,** and in the present example would proceed as follows. A first (simple) regression would be run to provide a slope that showed the impact of satisfaction on well-being: the first and last variables in the chain. A second (multiple) regression would then be run to provide the partial slopes that showed the impact of satisfaction and positive affect on well-being. If it is true that satisfaction can only have an impact on well-being through positive affect (total mediation), then breaking the chain by holding positive affect constant should remove any relationship between satisfaction and well-being. From a statistical standpoint, this means that we would expect a significant positive slope for satisfaction in the first regression, but a nonsignificant partial slope in the second regression when positive affect is held constant.

There are many other issues surrounding the analysis of mediation effects, some of which we will pursue in later chapters. The main purpose of this discussion is to introduce the notion of sequential regression and to give an initial sense of how it might be used to analyze mediation relationship patterns. It is worth noting that in practice sequential regression is also used to analyze moderation or interaction effects. In the earlier example a multiple regression would first be run to examine the independent effects of positive affect and satisfaction on well-being: the so-called **main effects.** Then a second multiple regression would be run that included the main effects variables and the interaction variable. Again our interest would be in comparing the results of the two regressions, this time in terms of R^2. Specifically, we would want to know whether there was a noteworthy increase in R^2 when the interaction variable was added into the regression equation. This increase can itself be tested using an F test. If it were significant, we would have evidence that there was an interaction effect of the two independent variables over and above their separate effect. As these two brief examples show, sequential regression provides a powerful tool for the analysis of complex patterns.

3.2.3 Generalizing the Composite Variable

Multiple regression involves expressing the relationship between one dependent variable (DV) and multiple independent variables (IV1, IV2, etc.) into various additive components. The components can be written:

DV = [slope 1(IV1) + slope 2(IV2) + ...
+ Y intercept]
+ residual

Slopes 1 and 2 represent the effects of IV1 and IV2, respectively, on the dependent variable. The dotted line indicates that we can add more independent variables, separately and/or in combination. The terms within the square brackets form a composite variable and are combined as shown to generate the predicted dependent variable score. Within this composite, the Y intercept can be seen just as an adjustment to ensure that the predicted and actual dependent variables are aligned by having the same mean. Finally, the residual captures that part of the dependent variable that is not predicted by the composite variable.

So far all of these multivariate ideas have been expressed in terms of multiple regression, a particular multivariate technique. It will be helpful now to reexpress the equation above in more general terms so that we will be able to use it each time we examine a particular technique in Part 2. This common framework will do more than any other device to help us make sense of the unity of multivariate techniques. The re-expression involves no more than the replacement of some of the words in the equation as follows:

DV = [coefficient 1(effect 1) + coefficient 2(effect 2) + ...
+ constant]
+ residual

The IV terms have been replaced by the more general term "effect." An effect may be a single variable, a so-called main effect; or it may be two or more variables in combination—an interaction effect. What were slopes are now referred to as **coefficients,** which is just a general instruction to multiply the variable value that follows by the coefficient value. Coefficients are also referred to as weights, so, for example, regression slopes may be described as regression coefficients or regression weights. The composite variable in the square brackets is often described as a weighted linear sum since it involves summing a set of variable scores, each of which has been weighted by a multiplier or coefficient. Rather than "weight," though, we will adopt the term "coefficient" since it is the most commonly used across multivariate techniques. The term "Y intercept" is specific to regression so it has been replaced by the more generic term "**constant.**" This word is also useful because it carries the connotation of a fixed adjustment. The terms "residual" and "error" are used synonymously. However, since error also has many other meanings, using the more precise "residual" is preferable.

With these terms in place, we can now reflect on some aspects of this cornerstone of multivariate analysis. The composite variable is often seen as a device for building a statistical "model" that accounts for patterns in data, such as the differences found in a dependent variable. As we have seen, this modeling can be viewed from the perspective of the individual—what is the predicted value on the dependent variable for this person; of individual differences—how much of the individual differences in the dependent variable can be explained by differences on the independent variables; and of group differences—how are group differences on the dependent variable related to group differences on the independent variables? The term "modeling" is helpful because it reminds us that it is up to the analyst to choose which effects should be included in the model. The statistical framework allows any number of variables in principle, singly or in combination, and in their original or some transformed state, so the choice can be immense. If the analyst is to account for differences in parsimonious and helpful ways, many effects have to be excluded, but this must be achieved without loss of important information. Well-developed theories should play an important role in guiding these choices. The data analyses themselves, though, also have a role to play by allowing the analyst to *compare* different models of the same data set. This model-comparison approach is one we will encounter repeatedly in Part 2.

Although the composite variable will provide a common framework within which we can locate all of the multivariate techniques in Part 2, we will find that the way in which coefficients are generated will differ across techniques. As we saw in Chapter 1, regression coefficients are calculated according to the principle of least squares. This means that coefficient values are found that minimize prediction error, that is, the sum of squares of the residuals. Another equivalent way to express the least square criterion is that it produces coefficient values that maximize the correlation between actual and predicted scores on the dependent variable. Each time we encounter a new multivariate technique, we will ask what criterion is used to generate the coefficients. The criterion will always be in the form of minimizing or maximizing something, but the something will vary by technique.

This completes our introduction to the core ideas of multivariate analysis. The composite variable, made up of a sum of weighted variables, is clearly a powerful tool with the potential to provide answers to a wide range of complex research questions. However well an analysis has been executed, though, the problem remains that the answers may be misinterpreted. In Part 2 particular issues surrounding the correct interpretation of results from specific techniques will loom large. In addition to these issues, there are what could be called "generic" misinterpretations, in the sense that they arise often and with respect to all multivariate techniques. In the following section we discuss the nature of these generic misinterpretations as a final preliminary before turning to the techniques themselves.

3.3 COMMON MISINTERPRETATIONS OF MULTIVARIATE ANALYSES

For our purposes, generic misinterpretations can be organized around four questions:

- What does "accounting for" differences mean?
- To whom or what do the results apply?
- What do the results of hypothesis tests mean?
- What does statistical control actually achieve?

The first three questions are not specific to multivariate analyses in that they can be raised for simpler forms of analysis as well. However, they are so fundamental that they deserve discussion in any introduction to data analysis. Moreover, the sophisticated and complex appearance of multivariate analyses can mislead the unwary into believing that the problems to which these questions point have been solved in some magical statistical way. In fact, multivariate analysis only helps to deal with the issues surrounding the last question and then only in a limited fashion, as we will see.

Answering these four questions leads quickly into deep statistical and philosophical waters. The aim of this section is not to dive into these waters, but to give some general sense of the ways in which misinterpretations can arise, so that they can be avoided. The references at the end of the chapter provide lively and accessible treatments of the underlying issues for those who wish to explore further.

3.3.1 What Does "Accounting for" Differences Mean?

The vague expression "accounting for" appears throughout this book and was chosen because it carries fewer potentially misleading connotations than many of the synonyms that are used in the language of data analysis. As we have seen, data analysts talk in terms such as *predicting* scores on, and *explaining* variance in, the dependent variable. It is also common to define the regression slope as the amount of *change* in the dependent variable when the independent variable changes by one unit. So it appears that analyses are able to provide predictions, explanations, and verification of causal processes. However, data analyses in and of themselves provide none of these.

The term "prediction" is a slippery one that has at least three meanings in the research context. In its strongest sense it means making a claim about what will happen in the future, and doing so successfully is seen as a mark of scientific progress. In another sense prediction is a synonym for constructing hypotheses about differences or relationships, without any reference to time. In the final meaning a predictor is just a statistical synonym for an independent variable. The presence of a "predictor" in an analysis suggests that the analyst believes that it may help to account for differences in the dependent variable and usually implies a formal hypothesis to that effect. However, even when it is shown that the "predictor" is significantly related to the dependent variable, this does not demonstrate any predictive power in the strong sense. The only basis for this interpretation would be if the data had been gathered longitudinally, where the independent variable is measured at one time and the dependent variable some time later. So interpreting "accounting for" in terms of prediction is justified by the design and analysis of a study, not by the analysis itself.

The term "explanation" is much more slippery and multidimensional than "prediction," so we will focus more on what it is not than on what it is. The first thing to note is that it is different from prediction. It is possible to make successful predictions without being able to explain why these predictions work. Similarly, the workings of a phenomenon may be well explained, but predicting its future states may be impossible because of the many other factors that enable or prevent the occurrence of these states. Turning back to data analyses, if we detect "explained" variance in a dependent variable it simply means that the independent and dependent variables covary in some systematic way. Why this covariation occurs remains open to explanation. Explanation lies in the realm of theory, not data analysis. Good theories generate testable hypotheses, and the results from the subsequent data analyses may be consistent with the hypotheses or not. There is clearly an important connection between theory and data analysis, but one should not be confused with the other. Moreover, this connection usually comprises a series of assumptions about measurement, sampling, and so forth, any of which may be faulty. So theory and data analysis not only are separate, but also have a loose linkage. All of this indicates that whatever "accounting for" means in data analysis, it is not equivalent to explanation.

A special and valued type of scientific explanation is that which provides understandings of causal processes. Once again, the fact that some terms in data analysis appear to refer to causal processes does not mean that statistical results, however sophisticated, can provide evidence of causation in and of themselves. Such evidence requires a reliance on theory, research design, and data analyses, tied together with explicit arguments. The most compelling evidence typically comes from a combination of an elaborated theory from which clear causal hypotheses can be rigorously deduced, a strong experimental design that mimics the productive or generative aspect of causation and that ensures tight control of confounding variables, and an appropriate analysis for the type of data that is generated. These are strong requirements that clearly cannot be replaced by data analyses alone, however complex they may be.

Other problems may also undermine attempts to treat statistical results as explanations of causal processes. Without a compelling theoretical explanation, a potential cause may be confused with a *marker* of that cause. As a facetious example, having white hair is strongly related to the incidence of many diseases, but no one suggests hair dye as an intervention to avoid these diseases. Hair color is simply a marker for age, which *is* causally implicated in the occurrence of disease. Statistical analyses are blind to the distinction between cause and marker, so other, additional ways must be found to develop accurate causal accounts.

Another problem in the statistical search for causes is that even complex statistical techniques are only designed to detect patterns consistent with what Lieberson (1985) calls symmetric or reversible causation. Referring to an independent variable or cause as X and a dependent variable or effect as Y will help to clarify this idea. Symmetric or reversible causation assumes that the amount of increase in Y produced by a unit increase in X will be the same as the amount of decrease in Y consequent on a unit decrease in X. But this symmetric behavior of increases and decreases may not hold. Once achieved, an increase in Y due to X may be irreversible whatever subsequently happens to X. Or a decrease in X may produce a partial reversal where Y returns to an intermediate value. Lieberson provides convincing examples of behavioral and social phenomena that display asymmetric causal processes and teases out the disturbing consequences for the study of such phenomena. For present purposes, the noteworthy consequence is that some causal processes cannot be captured by conventional data analyses. Most forms of analysis assume a billiard-ball-type of causation, but this is only one form of many. Again we arrive at the conclusion that while data analyses may contribute to the production of causal explanations under some circumstances, they cannot do so in isolation from other research activities, especially the activity of theorizing.

To complete this subsection, an even more fundamental issue about causality may be raised. The view of causal process served by conventional data analyses locates the process in the relations between variables. On this view, causal propositions are tested by evaluating the presence/absence, magnitude, and direction of relationships among variables. This is a task to which multivariate analysis is well suited, as we have seen. However, there are other views of causation that do not fit so well. For example, the realist conception locates causal forces in agents, not in relationships among variables (Sayer, 1992). So a realist causal account will explain how the causal powers of agents bring about changes. Further, such an account will refer to ways in which these causal powers may be enabled or constrained. A fundamental consequence of enablements and constraints is that a causal process will not necessarily be evident in any consistent way in patterns of relationships. Sometimes you see them, sometimes you don't. Deep waters are close by, so we will not pursue this further. The general point to note is that the causal accounts to which multivariate analyses can contribute are but one way to develop scientific explanations. So, to interpret their results meaningfully, the analyst has to buy into and, if necessary, defend a particular conception of causation, which itself is problematic.

3.3.2 To Whom or What Do Statistical Results Apply?

At first glance, this question appears to refer to the issue of generalizing beyond the data available. However, this is not the focus of this subsection. The issue here is that of being clear about what "units" are the object of an analysis at any given point. The potential problem of misinterpretation this raises is that of applying results to the wrong unit of analysis. In the analyses conducted in Part 1, we have consistently distinguished three "levels" of analysis: the individual, individual differences, and group differences. Analysts in different disciplines focus on different levels in this sense. A sociologist, for example, might conduct an analysis containing communities that aggregate into towns or cities. But in any statistical analysis there will always be multiple levels, whatever they may be, and the points to be raised here apply regardless of the nature of those levels or units. For the sake of consistency, the discussion will continue to be framed in terms of individuals and groups.

The most fundamental point to note is that since statistical analyses typically aggregate across individuals, the results do not refer or apply to any single individual. As we have seen, aggregate statistics, such as regression slopes, may be used to generate a prediction about an individual, but they do not quantify any attribute of any particular individual. This may seem self-evident, and most analysts are aware of it. However, there is a subtle version of this slide from the aggregate to the individual that is common, at least in psychology. Valsiner (1986b) has provided an intriguing demonstration of how even experienced researchers misinterpret simple correlations. The description of the correlation typically begins by referring to groups or an averaged relationship, but it quickly slides into a discourse about an idealized individual. This can be partly explained by the tension in psychology between a disciplinary focus on the individual and a general reliance on aggregate data to help us understand the individual. But Valsiner argues convincingly that many other cognitive processes lie behind the misinterpretation, and there is no reason to believe that these processes afflict only psychologists.

Once pointed out, the potential for interpretive confusion between group and individual is clear though not always easily avoided. The distinction between group differences and aggregated individual differences is harder to grasp but important because it provides another common cause for misinterpretation. The distinction can be sharpened up by considering again two bivariate statistics that refer to group differences and individual differences, respectively: the slope and r^2. A regression slope indicates how *group* differences on the dependent variable are related to *group* differences on the independent variable, where the groups are defined by values of the independent variable. In contrast, the r^2 statistic indicates how *individual* differences in a dependent variable are related to *individual* differences in an independent variable. So while the slope can be interpreted as the averaged effect of the independent variable on the dependent variable, r^2 cannot be interpreted in this way. The reason for highlighting this is that many research reports express a theoretical interest in group effects but then give pride of place to r^2-type statistics. This has the unfortunate effect of implicitly shifting research objectives away from examining group effects to maximizing the capture of individual differences on the dependent variable. As we will see in Chapter 4, since r^2 is highly dependent on variances and therefore unstable across samples, it has very limited uses. But the more fundamental concern is that focusing on r^2 at the expense of the slope can distort research objectives and lead to misinterpretations.

3.3.3 What Do the Results of Hypothesis Tests Mean?

In Chapter 2 null hypothesis testing was introduced as the predominant method used by social scientists to evaluate the role of chance in their results. Statistical significance with an alpha of no more than 5% is usually the license required for results to be treated as worthy of interpretation. This approach to setting a threshold for interpretability, based on a rejection that the pattern of results is due to chance, continues to play an important role throughout the realm of multivariate analysis, as we will see in Part 2. Over many decades methodologists have debated the meaning and worth of null hypothesis testing, usually from a critical perspective. Of all the procedures explored in this book, null hypothesis testing is the one that has received the most critical attacks. Yet despite this, it remains a cornerstone of data analysis in the mainstream social sciences. So it is important for any data analyst or user to gain some sense of the criticisms and to find a comfortable personal position.

Frank Schmidt, a very distinguished methodologist in psychology, has provided a memorable thumbnail sketch of null hypothesis testing that certainly pulls no punches. He wrote:

> If we were clairvoyant and could enter the mind of a typical researcher, we might eavesdrop on the following thoughts:
>
> > . . . If my findings are not significant, then I know that they probably just occurred by chance and that the true difference [or relationship] is probably zero. If the result is significant, then I know I have a reliable finding. The *p* values from the significance tests tell me whether the relationships in my data are large enough to be important or not. I can also determine from the *p* value what the chances are that these findings would replicate if I conducted a new study.
>
> Every one of these thoughts about the benefits of significance testing is false.

—Schmidt (1996, p.126)

This arresting paragraph with its knockout punch line raises a host of issues too complex to pursue here in any depth. A few comments, though, may help to explain Schmidt's conclusion. The outcome of a null hypothesis test turns on the *p* value, and it is misinterpretations of this probability that lay the foundations for further misconceptions. As we noted in Chapter 2, the *p* value represents the probability of finding a sample difference or relationship at least as large as that calculated *if the null hypothesis were true.* The italicized words highlight two important points. First, the probability is a *conditional* probability, not simply the probability of a particular sample value occurring at all. Second, the probability refers to the sample value, not to the hypothesis under test. The probability of a sample value conditional on a hypothesis being true is different from the probability of a hypothesis being true conditional on a sample value. So the *p* value tells us nothing directly about the probable truth of the hypothesis. The plot thickens further when it is appreciated that the null hypothesis usually under test, what Cohen (1994) calls the "nil hypothesis," is usually if not always false as a matter of fact. The notion that a particular population difference or relationship is precisely zero seems an odd assumption, and yet it provides the precisely defined start and end point for the testing process. Finally, if there is no probabilistic basis for the sample, either by probability sampling or random assignment, any inferences from the *p* value lack a clear point of reference. All of this suggests that while a chance interpretation based on the *p* value may be the most viable of those listed by Schmidt, the exact nature of that interpretation is not straightforward or always clear.

Schmidt also dismissed the claims that statistical significance implies reliability and replicability of results. This can again be appreciated as a consequence of the abstract nature and origin of the *p* value. In null hypothesis testing, the sample value is seen as one of an infinite set of possible sample values. This set of values has a frequency distribution—the sampling distribution—that defines the relative frequency or probability of any given value occurring. It is by reference to a particular sampling distribution, representing the null hypothesis, that the *p* value is derived. Given this highly abstract framework, it is hard to see how any implications about the consistency of future sample values could be drawn. The reliability and replicability of results can be based only on the cross-validation provided by repeated analyses, either on subsamples within the same study or on samples from different studies. The latter option has been greatly enhanced in recent years by the advent of meta-analysis, whereby results from different studies can be statistically amalgamated in a rigorous fashion.

The remaining claim, that statistical significance has implications for the importance of a result, is probably the most widespread misinterpretation, despite repeated warnings in methods texts. Even if we assume that a chance interpretation of statistical significance is defensible, it still has no direct implications for other sorts of significance—theoretical, practical, or otherwise. One way to see this is first to remember that the *p* value can be interpreted as the probability of committing a Type I error—rejecting the null hypothesis when it is actually true. Now add to this a further point from Chapter 2, that an effective way to reduce Type I error is to increase the sample size. This means then that sample size is one of the determinants of the *p* value. This is borne out in practice in large sample surveys in which even tiny differences and relationships turn out to be statistically significant. At best, statistical significance may be seen as a limited indicator that a result is unlikely to be due to sampling error and that it is worthy of interpretation. It is this *subsequent* interpretation, using criteria outside the analysis, which forms a basis for claims of importance.

The cumulative effect of these concerns has led an increasing number of methodologists to recommend the abandonment of null hypothesis testing. Schmidt (1996) advocates this not only on the grounds of the indefensibility of the procedure in his view, but also because of the damage it has done. He believes that progress, at least in psychology, has been retarded by widespread Type II error. In other words, many hypotheses and their parent theories have been wrongly dismissed on the statistical basis that the null hypothesis had to be accepted, when in fact it was probably false. The most common recommended alternative is to base chance interpretations of results on point and interval estimation, as discussed in Subsection 2.3.2 of Chapter 2. It is further recommended that replication interpretations should be based on cross-validations within studies and **meta-analyses** across studies. The latter are statistical methods for combining results across studies in order to evaluate the size and reliability of effects. Issues of substantive interpretation or importance should remain outside the domain of statistics. The meta-analysis recommendation has clearly taken hold as such analyses are now a commonplace in the research literature. However, chance interpretations continue to be based on null hypothesis testing in the main—reason enough to gain an understanding of its nature and limitations.

3.3.4 What Does Statistical Control Actually Achieve?

As we noted earlier, all of the preceding interpretive problems can arise in almost any analysis, and multivariate analyses are certainly not exempt. Finally, we turn to an issue that only arises when the relationships among three or more variables are being analyzed: the issue of statistical control. A naive interpretation of results that have been statistically controlled would suggest that the results must be somehow definitive or "correct" since they have been "corrected." In fact, there are various ways in which the exercise of statistical control may lead to distortions or, at best, limitations.

The first problem for the researcher is to decide *which* variables might be confounds and should therefore be included in addition to the chosen independent and dependent variables. In principle, the list of possible confounds is infinite, especially when we start to think about what variables might be confounded with confounds! The threat of the missing confound, and consequent distortion due to undercontrol, is ever present. An understandable response to this problem is to be overinclusive and control for a gigantic set of possible confounds. Unfortunately, this not only reduces the statistical power of the analysis, making it even more hungry for cases, but also runs the risk of distortion due to overcontrol. The very effects being sought may be obliterated or distorted by the complex adjustments required by the presence of numerous possible confounds. Once again we arrive at the conclusion that analyses need to be designed with reference to external sources, particularly theory and existing evidence. Moreover, as Cohen (1990, pp. 1304–1305) has observed: generally speaking in data analysis "less is more" and "simple is better." There is no easy answer to the question of which variables should be controlled in an analysis, and the results are always contingent on which choices have been made. Interpretations of results have to always keep their contingent nature in mind; in no sense can they be definitive.

The second problem is one we have already encountered in Chapter 2—that of measurement quality. Whatever control variables, i.e., potential confounds, are included, they must be measured with adequate reliability and validity. If they are not, the statistical control process will be undermined and may produce unpredictable distortions of the relationships between the independent and dependent variables. It is very easy to treat control variables as second-class citizens not deserving of the measurement efforts expended on the "real" variables. However, the consequences of such neglect may be disastrous, especially since the random error produced by the unreliable measurement of a single variable can ripple through the network of relationships in a multivariate analysis, causing widespread contamination. All of this means that any report of an analysis should include evidence to reassure the reader that interpretations of the results are not threatened by poor measurement of *any* of the variables.

The final problem concerns the consequences of statistical control for the interpretation of variables themselves. As we saw in Section 3.2, statistical control involves adjusting an independent variable by removing the variance it shares with a possible confound. So the original independent variable is replaced in the analysis by an adjusted or residualized version. This raises two interpretive issues. The first is that the nature of this adjusted variable may be different according to which other variables have been adjusted for. Accordingly, the results of two different analyses focusing on a particular independent variable may not be comparable if the set of control variables is not the same in each analysis. The second, deeper issue is the question of how an adjusted variable is to be interpreted at all. If the variable of positive affect, say, has been adjusted for satisfaction, what exactly is left? It is tempting to resort to some notion of uncovering the essence of a variable by stripping away those features it somehow shares with other variables. But given the inherent interrelatedness of so many variables, this does not seem very convincing.

Once again, we are led to appreciate the contingent nature of multivariate results and the cautious interpretations that are therefore required.

After this catalog of sticky problems, it may be tempting to stop reading and give up on multivariate analysis, perhaps even on simpler analyses. But it is important to reiterate that Section 3.3 has been all about general *limitations* on interpretations of results from multivariate analyses. It is not an attack on multivariate techniques themselves, but an attempt to encourage a critical and balanced approach to them. Every technique can be misused, and the more complex the technique, the more scope for unwitting misuse.

3.4 FURTHER READING

A good general orientation to the multivariate perspective can be found in Cohen, Cohen, West, & Aiken (2003, Chapter 1), while Darlington (1990, Chapters 1 and 4) provides an excellent discussion of the nature of confounding and statistical control. Baron and Kenny's (1986) discussion of moderating and mediating relationships has become the classic reference for this topic. The texts by Lieberson (1985) and Sayer (1992, especially Chapter 6) provide accessible critical accounts of some of the conceptual limitations of multivariate analysis. Tacq (1997) pays extensive attention to the question of how multivariate analytic frameworks map onto the structure of research problems. Runkel (1990) and Valsiner (1986a) examine in depth the issue of the relationship between single case and aggregate analyses. Good critical accounts of null hypothesis testing are available in Cohen (1994) and Schmidt (1996), and a contrasting view can be found in Frick (1996).

PART II

THE TECHNIQUES

FOUR

MULTIPLE REGRESSION

Multiple regression is a data analysis technique that enables the analyst to examine patterns of relationships between multiple independent variables and a single dependent variable. In its most basic form, the analysis requires that all variables have been measured on an interval scale. However, as we will see, the technique can be elaborated to accommodate independent variables that are measured on other types of scale and that may be rendered in more complex forms, singly and in combination. This flexibility makes multiple regression a highly popular technique that can be seen as a general form of analysis on which many other multivariate techniques are variants. This is why it takes pride of place in Part 2, both in terms of its location and relatively detailed treatment. An understanding of the nature and potential of multiple regression provides an excellent starting point from which other multivariate techniques may be explored.

Section 4.1 establishes the routine we will follow at the beginning of most chapters in Part 2. Here we return to the generic framework within which we can examine multiple relationships using any multivariate technique that centers on the weighted composite variable introduced in Chapter 3. Then we review the particular form this takes for multiple regression and the specific criteria by which this version of the composite variable is generated. This can be accomplished quickly because of our earlier discussions of regression in Chapters 1 and 3. Section 4.2 introduces several new statistics found in multiple regression and shows their use with examples from the research literature on subjective well-being. In Section 4.3 we focus on the details of a particular study to see how the issues of trustworthiness of results that we explored in Chapter 2 are treated in the context of multiple regression. Since most of these issues and their treatment are common to many multivariate techniques, the discussion in this section will provide a source of information that can be accessed repeatedly from later chapters and help to avoid unnecessary repetition.

Section 4.4 explores some of the common ways in which multiple regression can be elaborated to accommodate various types of independent variable. Then, in Section 4.5 we examine another feature of multiple regression's flexibility: the use of *series* of regression analyses to capture the complex patterns of moderating and mediating relationships that we encountered in Chapter 3. This section also discusses another type of sequential regression strategy in which the sequence of steps is driven by statistical criteria rather than the causal patterns supplied by the analyst.

4.1 THE COMPOSITE VARIABLE IN MULTIPLE REGRESSION

In Chapter 3 we introduced the key multivariate strategy of combining multiple variables into a composite. If the objective is to use the multiple independent variables (IV) to account for differences on a single dependent variable (DV), they can be combined as shown in the square bracket in the following general equation:

DV = [coefficient 1(effect 1) + coefficient 2(effect 2) + ...
+ constant]
+ residual

This says that the differences on a dependent variable can be decomposed into those that can be accounted for by a set of "effects" chosen by the analyst, and those that are not accounted for—the residual. Looking inside the composite, we see that an effect may be represented by a single independent variable or by a combination of them. Each effect has a coefficient or weight that can be interpreted as its contribution to the dependent variables when all other effects are statistically controlled. When an individual case's scores on the independent variables are multiplied by the appropriate coefficients, then summed and adjusted with a constant value, a new composite score is generated. This can be interpreted as a "predicted" score on the dependent variable for that case. Since the predicted and actual scores on the dependent variable will rarely coincide, there will also be a "gap" or residual: the actual score minus the predicted score. So for each case, the composite variable can be used to generate a predicted score and a residual that add up to their score on the dependent variable. These predicted and residual scores can be aggregated across cases to perform the analyses that lie at the heart of multivariate analyses.

Different types of multivariate analyses use different labels for the coefficients and sometimes for the constant in this generic equation. In the case of multiple regression, the equation becomes:

DV = [slope 1(IV1) + slope 2(IV2) + ...
+ Y intercept]
+ residual

Since the dependent variable is conventionally labeled as the Y variable in regression, the score generated by the composite in the square bracket is known as the **predicted Y score.** Clearly, the scores on this variable hinge on the values calculated for the slopes or regression coefficients. Where then do these values come from? When we explored simple regression, we looked at the slope calculation in some detail. But in the multivariate context we will ignore calculation strategies as such and instead focus on the criterion or criteria that guide the calculations. These criteria vary across different multivariate techniques. For multiple regression, as we previewed in Chapter 1 when discussing simple regression, the criterion is that of **ordinary least squares.** This can be interpreted in two complementary ways, referring to the predicted Y scores and residuals, respectively. The slope and constant values are chosen such that the correlation between the actual and predicted scores on the dependent variable is at its maximum. Alternatively, and reflecting the least squares label, the values are chosen so that the sum of squares of the residuals is at its minimum. The term "ordinary" simply means that the residual values are left in their original form, in contrast to some other more complex strategies that we will encounter later. So, in multiple regression the slopes and the constant are chosen to maximize the predictive power of the independent variables or, conversely, to minimize errors of prediction.

4.2 STANDARD MULTIPLE REGRESSION IN ACTION

A multiple regression analysis is referred to as "standard," "all-in," "simultaneous," or "direct" if it is conducted in one step. For example, Hayes and

Joseph (2003) used this type of analysis to find out whether people with different personality attributes experience different levels of happiness. In one of their analyses they found that higher scores on the Oxford Happiness Inventory (the dependent variable) were associated with lower neuroticism and higher extraversion scores but were unrelated either to conscientiousness scores or to age and sex. In reporting these results they lean heavily on two common regression statistics that we have yet to encounter: **adjusted R^2** and the **beta coefficient.** We can use their results to introduce these workhorses of regression analysis.

Hayes and Joseph's analysis shows that the five independent variables—age, sex, neuroticism, extraversion, and conscientiousness—account for 50% of the variance in the dependent happiness variable. This is supported by an "adjusted" R^2 of .50. The squared multiple correlation between the predicted and actual Y scores we have encountered before in Chapter 3, but what is the adjustment? The adjustment is by way of a deflation to remove an inflationary bias in R^2. The sample value of R^2 consistently overestimates the population value, and the larger the number of independent variables and the smaller the sample size, the worse this bias becomes. Accordingly, regression computer programs routinely provide a deflated estimate of R^2 that has been adjusted for the number of independent variables and the sample size. Although the adjustment does not completely remove the inflationary bias, it is this version of R^2 that is usually reported as the best available estimate.

In the present analysis the five independent variables account for half of the variance in happiness. Put another way, the predicted happiness scores calculated from the composite variable are correlated about .71 (the square root of .50) with the actual happiness scores. This form of calculation assures us that any correlations among the independent variables have been statistically controlled, and so the multiple R of .71 is not inflated by overlapping contributions to explained variance. The multiple R and R^2 values have therefore been adjusted in two ways: to take account of biases due to the number of independent variables and the sample size; and to take account of correlations among the independent variables. So it is not surprising that regression analysts give pride of place to adjusted R^2 as a useful "overview" statistic to describe the relationship between a set of independent variables and a dependent variable. Nonetheless, we will see later that caution is still needed as it can easily be overinterpreted.

The set of independent variables appears to be related to happiness, but what is happening with respect to particular variables? In their discussion

Hayes and Joseph write: " . . . Extraversion was the best predictor of happiness as measured by the Oxford Happiness Inventory, followed by Neuroticism" (p. 726). Clearly, attention has now switched from the predicted Y variable to the slopes inside the composite, each of which represents the impact of a particular independent variable. But the writers are somehow managing to rank-order the impacts even though extraversion and neuroticism may be measured with different units. This is possible because they are referring, not to the slopes in their original form, but to what are called "standardized slopes," or **beta coefficients.** In general, if we want to compare variables that are measured using different units, we can do so by transforming the original scores on each into **standard scores.** This simply involves subtracting the sample mean from a score and dividing this deviation score by the sample standard deviation. The resulting standard or ***z* score** is now in standard deviation units and is therefore comparable with any other standard score. If a regression analysis is conducted on standard scores, the slopes are themselves standardized and are now referred to as beta coefficients. So the composite variable can be calculated either in its original form with unstandardized slopes or coefficients or in standardized form with betas as coefficients. Note that the use of standardized variables always produces a constant or Y intercept value of zero, so this adjustment disappears from the composite variable when it is in standardized form.

Hayes and Joseph report betas of .47 and –.28 for extraversion and neuroticism, respectively. The interpretation of a beta is the same as for an unstandardized slope except for the measurement unit. So the beta for extraversion says that for every 1 standard deviation increase in extraversion scores, there is a .47 standard deviation increase in happiness scores. Or in difference terms, groups who differ by 1 standard deviation on the extraversion measure differ by .47 of a standard deviation on the happiness measure. The negative sign for the neuroticism beta indicates that for every 1 standard deviation increase in neuroticism, there is a .28 *decrease* in happiness. So, in this sample, happier people tend to have higher extraversion scores but lower neuroticism scores. Since the betas are both expressed in standard deviation units, the authors have some grounds for the comparative claim that extraversion has a bigger impact than neuroticism on happiness. Further, the fact that betas in a multiple regression are *partial* slopes means that any correlation between extraversion and neuroticism scores has been statistically controlled.

Of the five independent variables under analysis, Hayes and Joseph conclude that two—extraversion and neuroticism—have trustworthy associations with happiness as measured by the Oxford Happiness Inventory. (Interestingly, the pattern of results is different for two other measures of happiness.) Their conclusion is based on t tests of the betas following the usual null hypothesis testing procedure. For the extraversion beta of .47, they report a t value of 5.17 with an associated p value of less than .01; for the neuroticism beta of −.28, the t value is -3.0 with a p value again less than .01. Using a conventional alpha of .05, they therefore reject the null hypothesis that the population beta is zero in each case, in favor of the nondirectional alternate hypothesis that the population beta is not zero. None of the t tests for the other three betas have a p value of .05 or less, so these betas are treated as effectively zero and therefore not worthy of interpretation.

As we saw in Chapter 3, it is usual in multiple regression to precede the separate beta tests with an F test of whether R^2 is zero in the population. This is an efficient approach since, if this single test of the composite variable does not produce a statistically significant result, no tests of the betas are required or justified. This F test of R^2 is not reported by Hayes and Joseph. However, given the size of the adjusted R^2 and, more important, the presence of significant betas, the F test would clearly have led to a rejection of the null hypothesis that the population R^2 is zero. In summary, Hayes and Joseph used a standard regression to examine the relationships of extraversion, neuroticism, conscientiousness, age, and sex with happiness scores on the Oxford Happiness Inventory. The standard regression strategy meant that they could estimate the overall relationship of the set of five independent variables, and the relationship of each, with the dependent variable, while controlling for correlations among the independent variables. The adjusted R^2 statistic indicated that the independent variables accounted for 50% of the variance in happiness, and we can assume that this is a statistically significant result with a $p < .05$. Testing the betas with t tests showed that only extraversion and neuroticism were individually related to happiness with a $p < .01$. Inspection of the betas suggested that extraversion had a stronger positive relationship, while neuroticism had a weaker negative relationship, with happiness.

To cement these fundamental ideas in place and to prepare for more detailed discussions, we can turn to a study of nurses conducted by Budge, Carryer, and Wood (2003). The focus of interest in their study was on how aspects of nurses' work environment affect their health. Various dimensions of health were assessed using the well-known SF-36 questionnaire. We will concentrate on the results for the mental health subscale, which includes items on the occurrence of positive emotions such as happiness and of negative emotions such as depression. The scoring scheme produces a possible range from 0 to 100, in which a higher score indicates better mental health. Three aspects of the nursing workplace were examined using the revised Nursing Work Index: the degree of autonomy for taking actions, the degree of control over the practice environment, and the quality of interactions with other health professionals. So the particular regression analysis on which we will focus has mental health as the dependent variable, and autonomy, control, and professional relations as the independent variables. Age was also included as a fourth independent variable because of its confounding potential.

Before the regression analysis, a bivariate correlation analysis showed that all four independent variables were positively and significantly ($p < .05$) related to mental health. However, since the three Nursing Work Index variables were themselves correlated, the possibility of confounding arises and thus the need for a controlled multivariate analysis. The standard regression for these variables produced an adjusted R^2 of .11, with an F value of 5.78, $p < .001$. These figures indicate that the four independent variables accounted for 11% of the variance in mental health and that this was unlikely to be due to chance. Or more precisely, this R^2 figure is inconsistent with the hypothesis that R^2 is zero in the population sampled. This is reassuring, but were all of the independent variables related to mental health when their intercorrelations were controlled?

The reported betas show that this was far from the case and that only the professional relations variable had a statistically significant relationship with mental health. The beta for this variable was .31, which was significant by t test at $p < .05$. The betas for autonomy and control were both .01, while that for age was .11, and all three were nonsignificant relative to the conventional alpha of .05. Although Budge et al. reported these betas, rather than providing the associated t and p values, they calculated a 95% confidence interval around each one. So the beta estimate of .31 for professional relations had a confidence interval lying between .13 and .49. This means that while the best point estimate of the population beta was .31, there was a 95% probability that it could be as low as .13 or as high as .49. Since this interval does not include zero, this is consistent with rejecting the null hypothesis with an alpha of .05. Similarly, since the intervals for all of the other betas did include zero, we can infer that none were statistically significant. As we noted in Chapter 2, many argue that confidence intervals are preferable to null hypothesis testing since they avoid some conceptual pitfalls and are more informative.

To complete this section on standard regression, it will be useful to introduce two further statistics: the **semipartial and partial correlation coefficients.** Regression analysts report them less commonly, but they provide further insight into the logic of correlation and regression. Budge et al. do not provide these statistics in their article so they have been calculated from the data that the authors have kindly provided, again focusing on the relationship between professional relations and mental health. The simple or "zero-order" correlation between these two variables of .35 captures their association without taking into account their relationships with any other variables. The semipartial (sometimes called part) correlation between these two variables is .25. This number indexes the same relationship, but now with statistical control of the correlations between professional relations and the other three independent variables. The professional relations variable has been residualized with respect to the other independent variables, and it is the remaining part of the variable that has been correlated with mental health. Since this is a correlation coefficient, it can be squared to obtain an explained variance statistic. The squared semipartial correlation is .063, which says that professional relations explains 6.3% of the variance in mental health over and above that explained by the other independent variables. It is the *unique* contribution that professional relations makes to explaining *all* of the variance in mental health.

The partial correlation coefficient for professional relations and mental health is .26. For this coefficient, *both* the professional relations and mental health variables have been residualized with respect to the other independent variables. Their partial correlation indexes their relationship when *all* other relationships have been controlled, including those between the other independent variables and the dependent variable. So a partial correlation controls for all of the relationships among the variables, while a semipartial controls only those among the independent variables. Again the partial correlation can be squared, and this gives a value of .068. The squared partial correlation indicates that professional relations makes a unique contribution of 6.8% to explaining the variance in mental health *that is left unexplained by the other independent variables.* Since the other variables explain hardly any variance in mental health, the semipartial and partial correlations are very similar in magnitude.

4.3 TRUSTWORTHINESS IN REGRESSION ANALYSIS

In Chapter 2 we discussed how the trustworthiness of results from any statistical analysis can be affected by problems of sampling, measurement, the role of chance, and the technical assumptions required by the chosen analytic technique. In this section we will revisit all of these issues but now with specific reference to multiple regression analysis. The Budge et al. study of nurses will remain the focus of attention throughout in order to avoid distractions. As noted earlier, having access to the original data means that we can dig more deeply into issues than is usually possible with published articles. The depth of discussion in this chapter will also enable shorter discussions in future chapters when the same issues recur.

4.3.1 Sampling and Measurement Issues

In the Budge et al. study data were collected from 225 nurses. However, once all cases with any missing data were excluded (so-called listwise deletion), 163 remained for the regression analyses. No attempt was made to replace missing data for individuals with imputed values such as the mean since this number of cases provided sufficient statistical power and met generally accepted criteria for adequate sample size in standard multiple regression. What then are these criteria? As Tabachnick and Fidell (2001) point out, this is a complex issue because adequacy depends on the chosen Type I and II error rates, the number of independent variables, the expected magnitude of relationships, the reliability of measurement, and the frequency distribution of the dependent variable. They provide helpful extracts from the comprehensive guide compiled by Green (1991), including two useful rules of thumb. If we assume conventional choices of .05 and .20 for Type I and II error rates, respectively, good reliability of measurement, and a normally distributed dependent variable, the minimum sample size for detecting a medium size R^2 is calculated as 50 plus 8 times the number of independent variables. In the present example this gives $50 + (8)(4) = 82$. The minimum sample size for detecting a medium-size beta under the same assumptions is calculated as 104 plus the number of independent variables = 108. Since the Budge et al. analysis examined both R^2 and betas, as is usually the case, the larger value of 108 was chosen as the minimum. The actual sample size of 163 therefore easily met this requirement, though it is important to note that choosing lower error rates, using unreliable measures, or finding a nonnormal distribution of the dependent variable would result in a need for a larger sample size than 108. In practice, none of these eventuated, as we will see, and the actual effect size also legitimated the sample-size calculation. It is also important to note that these sample-size calculations refer to standard regression only. If more complex regression strategies are used, such as those we will discuss in Section 4.5, the number of cases required increases substantially.

The sample size in the nurses' study is more than adequate to make the results trustworthy, but what of the nature of the sample and its manner of recruitment? Participants were sought from all of the 359 Registered Nurses who then worked at least 32 hours a week in a particular general hospital. Of these, 225 (62.6%) returned the questionnaire and 163 (45.5% of 359) provided a complete data set. Clearly, questions about the nursing workplace can best be answered by qualified nurses actively engaged in their work, a requirement well met by this sample. Further, the study has a well-defined population, all of whom were approached to participate. However, the actual sample in the analysis is potentially problematic, not because of the fact that it constitutes less than half of the population, but because it *may* be unrepresentative with respect to the study variables. The authors highlight the diversity of the sample characteristics, such as age, qualifications, and nursing specialties. This is reassuring in some ways but does not remove the possibility of sampling bias with respect to the work and health variables of central interest. Perhaps nurses with extreme work or health problems were less likely to devote time to completing the questionnaire. Or perhaps such nurses were more likely to respond, seeing the study as a way of drawing attention to their problems. We simply do not know if any of these response patterns occurred or if they affected the regression analysis.

These comments are not meant to be particularly critical of this nursing study but to highlight a common problem in statistical analyses. Even when a population has been defined, as in this study, the specter of sampling bias often places constraints on the interpretation of results. Moreover, nothing in the multivariate toolbox can be used to fix the problem. This means that the results of testing null hypotheses, and even of estimating confidence intervals, do not have a straightforward interpretation; the probabilistic inferences do not map onto the actual sample and population. As we noted in Chapter 3, the seriousness of this problem depends partly on how the analyst conceives of the role of statistical inference and of the relative merits of hypothesis testing and estimation. The authors of the nursing study are clearly aware of the advantages of confidence intervals and, at the time of writing, are conducting a replication study to test the replicability of their findings. In common with many other analysts, they are nonetheless continuing to use statistical inference as a way of evaluating to what extent their results might be due to chance, despite the absence of a probabilistic sampling framework.

Turning to measurement concerns, we can deal with these quickly since hardly any of the scaling and quality issues that we discussed generally in Chapter 2 are specific to regression analysis. With respect to scaling issues, regression requires that the dependent variable be measured on at least an interval scale. As we will see in a later section, more flexibility is possible for the scaling of independent variables, but it is clear that Budge et al. treated all of their variables as interval level. In common with most analysts, they offer no explicit justification for this. Instead they emphasize the well-established nature of their measures and imply that earlier users of the measures have treated their data as being on an interval scale without detriment. This may seem a little weak by way of justification, but it is common practice in the social sciences.

Budge et al. provide extensive information on the reliability of their measures, focusing on Cronbach's alpha as an index of internal consistency. They report alphas in excess of .7 for all of the Nursing Work Index subscales, based on results from their own study and from others. Even more reassuring are the alphas for the SF-36 subscales, which virtually all exceed .8 in their study and in a national New Zealand sample. The validity of the Nursing Work Index subscales is supported by reference to earlier research, which shows them to be "significant predictors of patient satisfaction, hospital mortality rates and nurse outcomes" (p. 262). No specific information is presented on the validity of the SF-36 subscales, but reference is made to an article on their construction that would typically provide such information. Although the information provided on measurement quality is not extensive, it is again not unusual in this respect. On the basis of what is provided, there seem to be no measurement grounds for doubting the trustworthiness of the regression results. The scaling, reliability, and validity of the measures used appears to be at least adequate.

4.3.2 Checking Assumptions in Multiple Regression

The legitimacy of a multiple regression analysis depends in part on how well assumptions have been met about the normality and variances of frequency distributions, the linearity of relationships, and the independence of cases. A general overview of these issues was provided in Chapter 2. Now we will explore their application in the context of multiple regression, using the Budge et al. data.

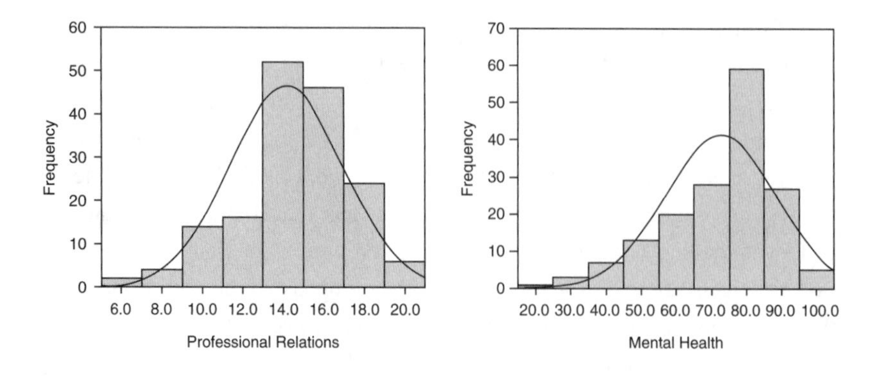

Figure 4.1 Histograms for the Professional Relations and Mental Health Variables

The **normality** of a variable's frequency distribution can be examined graphically by constructing a histogram and superimposing the shape it should follow if it were perfectly normal. Histograms for the professional relations and mental health variables appear in Figure 4.1. Unlike the frequency ladders used in Chapter 1, histograms show the values of the variable along the horizontal axis, and the frequencies on the vertical axis.

Neither distribution is normal: a perfectly symmetrical distribution around a single peak. But this is to be expected in real, messy, sample data as opposed to the idealized population distribution to which the normality assumption applies. There is some suggestion of a negative skew in the mental health distribution and a noticeable peak around 80 poking above the normal envelope. Statistical tests of normality are available but, given the robustness of the normality assumption, it is probably safe to conclude in this case that the two variables are sufficiently normal in their distributions without further exploration. We could now proceed to examine the histograms for the remaining independent variables. However, approaching normality in this way only takes us part way to meeting the particular requirement of multiple regression. This technique, like many other multivariate techniques, requires that the data be not just normal but *multivariate* normal. Before we look at this requirement more closely, it will be helpful to bring the issues of equality of variance and linearity into the picture. An initial sense of how well these assumptions have been met can be gained by examining Figure 4.2, which is a bivariate scatter plot of the relationship

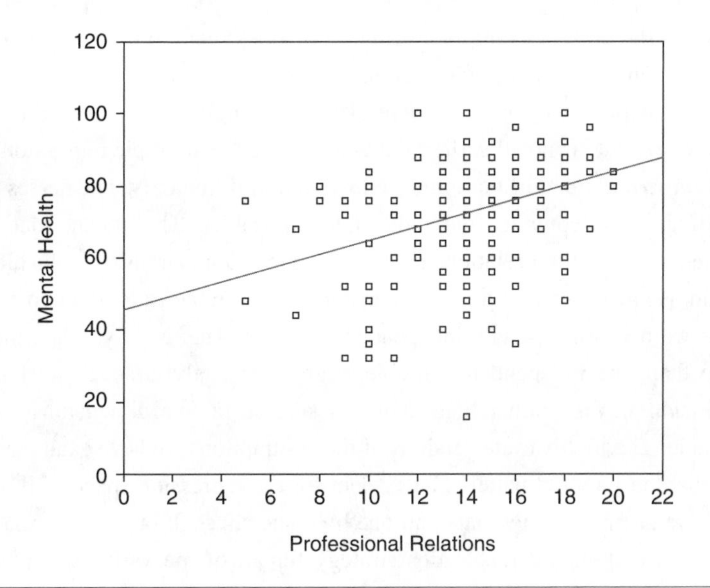

Figure 4.2 Scatter Plot of Professional Relations With Mental Health

between professional relations and mental health, complete with its simple regression line.

The variance assumption requires that the spread of data points either side of the regression line does not change markedly along the line, that is, the points should be enclosed more or less in a uniform band either side of the line. When this assumption is satisfied, the data are said to display homogeneity of variance or **homoscedasticity** (as opposed to **heteroscedasticity**). There is some slight suggestion in this example of a midline bulge, but this is mainly due to a single case at the bottom of the plot—a potential outlier—about which we will have more to say later. Overall, it seems safe to proceed on the assumption of homoscedasticity.

The **linearity** assumption requires that the shape of the data cloud be better summarized with a straight line than with any other type of line. It is important to appreciate that the assumption does not require the data to be *highly* linear, that is, to cluster tightly about a straight line. Whether or not this is the case has no implications for the legitimacy of a linear regression. Rather, the concern is that the data show no obvious signs of *non*linearity whereby a curved line of some description could be drawn through the center of the data cloud. In the present example a straight line does seem to be the best choice, and so the linearity assumption seems secure.

To this point we have considered variables singly (univariate) or as a pair (bivariate). But, as noted earlier, the assumptions for multiple regression refer to ***multivariate*** **normality, homoscedasticity, and linearity.** The nurses' multiple regression contains four independent variables. This means that if we wanted to depict their relationship with the dependent variable graphically, we would need to construct a five-dimensional space rather than the two dimensions we need for a single independent variable. This is why, when there is more than one independent variable, regression analysts start to refer to a regression *surface* rather than a line: a surface in five dimensions for our example. The multivariate versions of the assumptions we have examined dictate that the required patterns be evident for the regression space *as a whole.* We have taken two univariate and one bivariate slices through that space, and we could and should repeat this strategy for all of the variables. However, although this may seem counterintuitive, it is possible for all the slices to appear well behaved, but for their multidimensional aggregate still to be ill behaved. But how can we detect such a possibility, given the mind-blowing challenge of envisaging a five-dimensional space?

The answer is that we have already generated the necessary multivariate diagnostic tools by creating the composite variable and its accompanying residual variable. These two variables can be used to represent the patterns in the multivariate space that we need to explore. Engaging in this sort of multivariate diagnostic activity is known as **residual analysis.** The multivariate normality assumption can be examined by simply constructing a histogram of the residuals. The residual histogram for the nurses' multiple regression analysis is shown in Figure 4.3.

Residual scores are on the same scale as the dependent variable, the mental health scale in our example. However, in the histogram they appear in standardized form, that is, in standard deviation units. This is helpful in the present context because, by definition in a normal distribution, standardized scores lie between –3 and +3. So when residuals are presented in this form we can check the range and provisionally treat any cases with values outside these limits as outliers that might be distorting the analysis. We can now use the histogram as a tool to test the multivariate normality assumption in the knowledge that we are looking at a reflection of the distributions of all of the variables in the multidimensional space. Once again we can be generally reassured, but note that

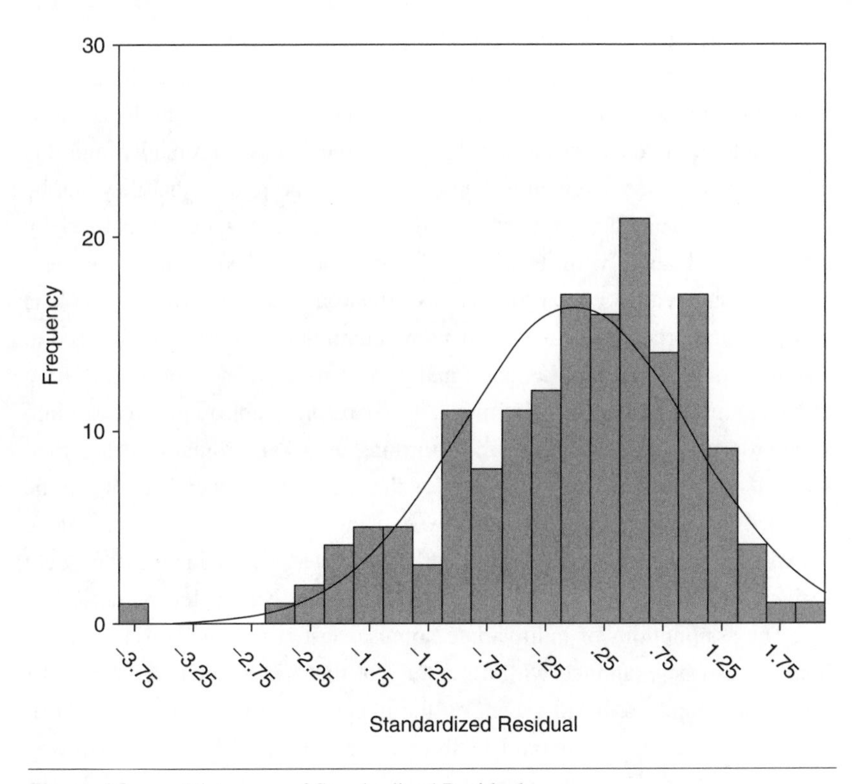

Figure 4.3 Histogram of Standardized Residuals

there is an outlier beyond the –3 boundary. It is important not to overreact in that only one case out of 163 meets this outlier criterion, and the case is less than one standard deviation beyond the limit. Nonetheless, the case was checked to see if this individual's data had been entered correctly, and if there was anything unusual in the data profile. Since the data seemed valid, the regression analysis was conducted again with this case omitted, to check its influence. Very similar results were obtained and so the published analysis included this case. This merely scratches the surface of how potentially influential cases such as outliers may be identified and dealt with. There are many other techniques that can be explored by consulting references given at the end of this chapter.

If the multivariate normality assumption seems to have been violated, what can be done? The first step is to track down which variable(s) is problematic by looking at univariate distributions. Then the solution depends on the particular way in which the distribution departs from normality. As we have seen, if outliers are a problem, they may be excluded from the analysis as long as the remaining sample is still large enough to meet the size requirement. If skewness is a problem, the distribution may be transformed toward normality, and this transformation may also help to pull outliers into the fold so that they can be retained. For example, moderate positive skewness may be normalized by replacing each score with its square root; or more marked positive skewness may be countered by a logarithmic transformation. If a transformed variable is used, it is important to appreciate that any interpretation of results will need to take this new scaling into account. Finally, if a variable is nonnormal by virtue of having more than one peak of mode, it may be preferable to group cases into an appropriate number of categories, forming either an ordinal or categorical scale. If the rescaled variable is one of the independent variables, it can be included in a multiple regression analysis using dummy coding—a technique we will introduce in the next section. If the rescaled variable is the dependent variable, it will be necessary to resort to another multivariate technique.

The assumptions of multivariate homoscedasticity (equal variances) and linearity can be examined with a scatter plot that has residual scores on the upright axis and predicted scores on the dependent variable along the horizontal axis. Again, it is helpful to show these two variables in standardized form, that is, they have a mean of zero and a standard deviation of 1. The scatter plot for the nurses' regression analysis is shown in Figure 4.4.

When we examined the variance and linearity assumptions from a bivariate perspective, we inspected the patterning around the regression line that represented the impact of professional relations on mental health. Since the predicted values in Figure 4.4 are generated by a composite of all of the independent variables, it would seem logical to plot the predicted scores against the dependent variable scores—a multivariate extension of Figure 4.2. This can be done and is informative, but even more informative is the present figure that plots predicted scores against residual scores. This, too, has a regression line that can be used as a reference point, but in this case the line is notional and superimposed rather than calculated. The line is horizontal and emanates from the zero residual value. This reflects the fact that the predicted scores and residuals should be independent of, or uncorrelated with, each other if the linearity assumption holds. Put another way, they represent explained and unexplained differences, respectively, that, because they are mutually exclusive, can be added together to represent all of the differences in the dependent variable.

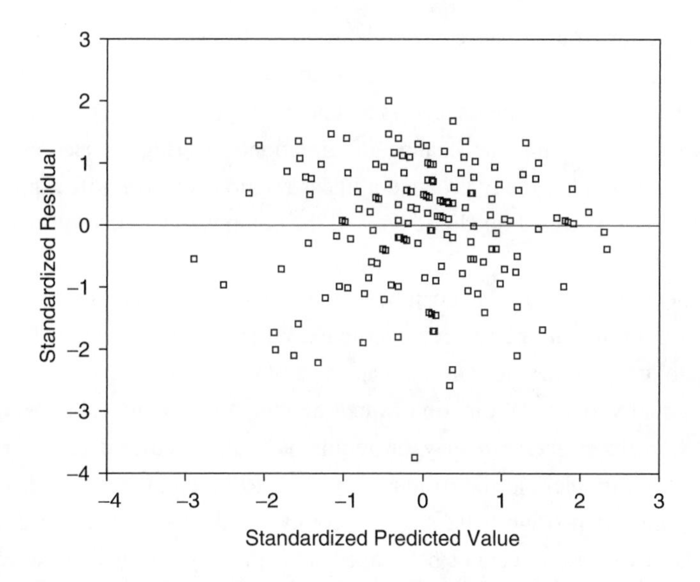

Figure 4.4 Scatter Plot of Predicted Y Values With Residuals

As in the bivariate context, the multivariate equality of variance assumption is met if the data points form a uniform band on either side of the regression line. Since we know that only the professional relations variable is related to mental health, it should come as no surprise that the bivariate and multivariate plots provide a similar message in this respect. The scatter plot can also be used to assess multivariate normality in terms of skewness and outliers, though the residual histogram provides a clearer picture. As before, the slight negative skew is evident in the asymmetry of the points, with a bigger spread below than above the line, and the outlier is again clear at the bottom of the plot. The multivariate linearity assumption also appears secure since there seems to be no underlying pattern in the data points that would be better summarized by a line that is not straight. This type of scatter plot is clearly a powerful diagnostic tool that can be used to evaluate a range of multivariate assumptions.

If the multivariate equality of variance assumption is violated, for example, with an increasing spread of residuals at higher levels of the predicted scores on the dependent variable, transformation of offending variables may again be a useful strategy. It is sensible to consider the normality assumption prior to this, as transforming a variable to normality often simultaneously solves an unequal variance problem. In other words, what appear to be two violations turn out to have a single source. Transformations can also be helpful in some situations where the assumption of linearity is violated. For example, if the scatter plot suggests an underlying pattern in the residuals that follows a curve with one change of direction, squaring the scores of the offending variable(s) may produce a linear relationship. We will explore this addition of a so-called quadratic term to the composite variable further in the next section.

There are limits to what can be achieved by transformations when violations of assumptions are severe or complex. When this occurs it may be appropriate to turn to types of regression analysis other than ordinary least squares. If the equal variance assumption cannot be met, a technique called **weighted least squares regression** may be required. If the predicted scores on the dependent variable and the residuals turn out to be correlated, another technique called **two-stage least squares** could be adopted. When the relationships in a regression cannot be coaxed into linearity, **nonlinear regression** techniques may be called on. With respect to scaling issues, if the dependent variable is clearly not defensible as being on an interval level scale, **ordinal regression** is an option. Or, if the dependent variable is best treated as categorical, **logistic regression** is an appropriate alternative, as we will see in Chapter 5.

The remaining assumption we have yet to explore in multiple regression concerns the **independence** of the cases. Now that we have introduced the notion of residual analysis, we can rephrase this assumption as requiring that the residual scores be independent of each other. This means that if we were to list the residual scores in the order they appear in the data set, there should be no discernible trend or patterning such as clusters. Inspection of the residuals for nurses' regression in this way revealed no obvious patterns. Nor was it possible to imagine ways in which nonindependence might have occurred given the way in which the data were collected. However, since the patterning can be hard to detect visually, it is also helpful to calculate something called the **Durbin-Watson statistic.** The more the value of this statistic deviates from 2, the more likely it is that the residuals are not independent. In the present example the obtained value was 1.71—sufficiently deviant to raise doubts. It is possible to test the statistical significance of this value, although the process is not straightforward and may have an inconclusive outcome (Cohen et al., 2003, p. 137). This was in fact the outcome for the present analysis, and so a question mark remains over whether the assumption of independence of residuals has been violated. The consequence of this violation is to inflate Type I error, and accordingly, Stevens (2002, p. 260) recommends reducing alpha as one way of coping with the problem. In the present case an alpha of .01 was adopted that produced no change in the pattern of statistically significant results.

Although not an assumption as such, one final issue deserves attention as we consider the legitimacy of multiple regression results. The problem is known as **multicollinearity,** which occurs when independent variables are highly correlated in some way. This has the effect of making it harder to reject null hypotheses about regression coefficients or, for estimation, to increase the confidence intervals around the coefficients. In the extreme case that there is a perfect correlation among the independent variables—a condition known as **singularity**—the regression analysis has no solution and will grind to a halt. Some sense of the independent variable relationships can be gained by examining their simple intercorrelations, which in the present case range between .48 and .69. However, it is also important to explore their multivariate relationships since there may be correlations among *combinations* of variables. This can be done with tools such as the **tolerance** and **variance inflation factor** statistics, which assess the multivariate association of each independent variable with the set of all of the others. None of these statistics suggested any high correlations among the independent variables. However, it is noteworthy that the highest simple correlation linking the independent variables with the dependent variable was .35 for professional relations, while the correlations among the independent variables ranged from .48 to .69. This type of pattern can make it difficult to evaluate the relative contributions of the independent variables since the amount of adjustment produced by the partialing process can overwhelm and distort the relationships of interest. In other words, although statistical control can help to reduce confounding, there are limits to what can be achieved when independent variables are more highly correlated with each other than with the dependent variable. Problems of multicollinearity may be amenable to another type of regression analysis called **ridge regression,** though it is one of the more contentious techniques in this area.

4.3.3 Other Issues of Trustworthiness

To complete this section on the trustworthiness of regression results, we will return to three issues that surfaced in Chapter 3 but that can now be given more focus. Although none are specific to the multiple regression context, their inclusion here will make Section 4.3 a useful overview that can be cited repeatedly in later chapters. The three issues are the effects of variability on covariability and thereby on replication, the effects of conducting multiple tests in a single analysis, and the importance of matching statistics to research questions. Failure to understand the implications of these issues can lead to misplaced trust in the findings of a particular statistical analysis.

In an earlier discussion on statistical control we noted that two variables can only covary if each one varies. If either variable has no variance and is therefore a constant, the two variables cannot be correlated. This is the extreme version of a more general point that the extent to which variables are correlated is constrained by their variances. Generally speaking, the more restricted the variance—sometimes called **restriction of range**—the lower the ceiling imposed on any correlations. This is an issue in its own right as it means that the failure to find an expected relationship of a particular magnitude may be due to restricted variance rather than to the true absence of a relationship. Here then is another reason to examine the univariate distributions of variables—to check for variability that is notably less than the measure allows. But there is a further implication concerning replication. The more statistics rely on variances or standard deviations, the less stable they are across samples because the variances themselves may well differ by sample, particularly if the sample sizes differ markedly. In terms of regression statistics, unstandardized slopes are more stable than betas or R^2 because the latter two, and any other correlational statistics, rely heavily on standard deviations for their standardized form. Ironically, this means that two of the most used regression statistics—R^2 and beta—are least likely to replicate across samples. It is important to note that the adjustment to R^2 that we discussed earlier does not address this issue, and nor does any aspect of the statistical control process. In fact, the more complex the control, the more scope there may be for instability. This is not an argument for discounting R^2 and betas, but it does highlight some of their limitations.

Turning to the second issue, we can see that by its very nature multivariate analysis tends to involve the use of multiple tests within one analysis. The use of omnibus tests can reduce the number, but if they indicate statistical significance, further tests inexorably follow. As we noted in an earlier discussion, the p values typically provided by most computer programs assume a "one-dip" logic, so that multiple dips increase the risk of Type I error. Various strategies have been suggested for dealing with this problem. One is what might be called the entry ticket approach. This proposes that multiple tests are justified and protected as long as they have been accessed through a significant multivariate test. So in multiple regression some analysts regard a significant R^2 as a justification for taking the p values of the slope tests at face value. Probably the most popular alternative strategy, in principle if not always in practice, is to make some adjustment to the p values of the slope tests by altering the alpha threshold. The simplest version of this, called **Bonferroni adjustment,** simply divides the conventional alpha by the number of tests to be conducted and then uses this as the required alpha. So if the conventional alpha of .05 were adopted and there were five tests to be conducted, an alpha of .01 would be applied to each test. This is just a nod in the direction of a very complex area, and more detail can be found in the references at the end of this chapter and in Section 6.1 of Chapter 6.

The third and final issue of trustworthiness concerns the match between research questions and the use of particular statistics. This is an issue we discussed briefly in Chapter 3, but which is sufficiently fundamental to bear repetition in the context of a particular multivariate analysis. We noted in Chapter 1 that analyses usually provide at least three perspectives: the individual case, differences between individual cases, and differences between groups of cases. A case may not be an individual, such as in an analysis of a sample of businesses, but the point about multiple perspectives still holds. As we have seen, in the regression context, slopes provide information about *group* differences, while R^2 and other correlational statistics index *individual* differences. None of these statistics in themselves provides information about the individual case. Clearly, information about a case may be extracted, such as their predicted score on the dependent variable. But the magnitude, sign, and statistical significance of regression and correlation coefficients have nothing to say about a given individual.

The reason for laboring this multiple perspectives point is that reports of statistical analyses sometimes display a mismatch between the perspectives implied by their research questions and those provided by their statistics. The nature and extent of this mismatch provides another basis on which the trustworthiness of the results may be questioned. Using statistics from the nurses' regression analysis to reach conclusions about the behavior and experience of an individual nurse is clearly untenable, whether this refers to a particular nurse in the sample or to an idealized individual nurse. This type of information can come only from studying individuals as such, even if the results are then aggregated. The authors of this article are clearly sensitive to this point and make no attempt to discuss their findings in terms of the typical nurse. As we noted in Chapter 3, this sensitivity is not always displayed in reports of statistical analyses.

Distinguishing between research questions about individual differences and those about group differences is more subtle and challenging. In the nursing study the writers express their interest in "correlates," "associations," and "links" between the workplace and nurses' health. Their results highlight the magnitude and the statistical significance of simple and multiple correlation coefficients, so everything is implicitly framed in terms of individual differences. The only apparent exception to this is their use of betas—standardized regression slopes. However, this is an oddly hybrid statistic in terms of the distinction between individual and group differences. From one perspective it conveys group differences in standardized form, but the standardization process turns it into a type of scale-free correlation coefficient. In fact, in simple regression the beta and simple correlation coefficient are identical. It is also noteworthy that the objectives of this study could have been achieved using part and partial correlation coefficients rather than betas. Since the significance test of these three statistics is the same, the messages would have been the same and would all have been in pure correlational, individual difference form.

It is striking that the writers do not include information on the unstandardized regression coefficients that would have conveyed the magnitude of the impact of professional relations on mental health, among other relationships. Even the betas are not interpreted in any way; they are just used to test the statistical significance of an effect under statistical control. As usual, this is not a criticism of the article, which is a good example of its type, but an attempt to highlight the exclusive emphasis on correlational, individual difference information in the data. The approach is internally consistent, but did the writers really have no interest in quantifying the extent of the impact of professional relationships in terms of *group* differences and in terms of the measures used? This seems like potentially useful information and, as we have noted, is likely to provide a less unstable picture across sample than that given by correlational statistics.

Many reports of statistical analyses are less consistent than our example. Often in the introduction interest is expressed in how an independent variable affects a dependent variable: a question that is best answered by regression coefficients, that is, *group* differences. But in the results section attention often shifts to R^2 and betas, typically to the magnitude of the former and the statistical significance of the latter. So what starts as a concern with group differences can drift into an individual difference account. This can lead to erroneous conclusions about the size of an effect as a noteworthy group difference can occur even though only a small percentage of the variance in the dependent variable is explained. For example, in the area of heart disease epidemiology, group differences in attributes such as blood cholesterol level may be linked to striking increases in heart disease risk for groups, while explaining small amounts of the individual differences in heart disease incidence. The trustworthiness of results as a basis for thought and action thus partly depends on ensuring that research questions and statistics match in terms of their level of analysis.

4.4 ACCOMMODATING OTHER TYPES OF INDEPENDENT VARIABLES

In their most basic form simple regression and multiple regression assume that any independent variable has been measured on at least an interval scale. Other types of independent variables can easily be accommodated, though, without too much effort. To exemplify some of the possibilities, we will continue with the nursing multiple regression example and specifically with the association between professional relations and mental health. To this point we have treated the professional relations variable as being on an interval scale. Now we will explore what happens if we assume that the variable was measured on a categorical or ordinal scale.

Imagine that we decide to categorize the professional relations scores into two categories by assigning those below the median into a "bad" group and those at the median or above into a "good" group. This is the weakest assumption we could make about the measurement properties of the professional relations measure, that it is only sufficient to divide respondents into the categories of a dichotomous categorical variable. If we assign a value of zero to the bad group and 1 to the good group, we have now created what is called a **dummy variable,** which can be entered into a regression analysis without further ado. What happens when we repeat the standard multiple regression in the continuing example, only with the professional relations variable replaced by the dummy variable?

In the original regression the unstandardized slope for professional relations was 1.7: Groups who are one unit apart anywhere on the professional relations scale differ by 1.7 units on the mental health scale when the other three independent variables are controlled. In the parallel dummy variable analysis, the unstandardized slope is calculated as 4.7. Since there are now only two groups we can simply say that they differ by 4.7 units on the mental health scale when the other independent variables are controlled. So the partial slope for a dummy variable is a *mean difference between two groups,* adjusted for any relationships between the dummy variable and the other independent variables, as usual. The positive value of the slope in this case says that the 1 (good) group has a higher mental health mean score than the 0 (bad) group; a negative sign would indicate a difference in the opposite direction. As usual, the slope can be expressed in standardized (beta) form and its significance evaluated with a t test. It is noteworthy that in this case the associated p value is .09, so in this dichotomized form professional relations has no impact on mental health. This illustrates nicely how reducing the differentiating power of a measure, and thereby the variance of the measured variable, may attenuate relationships. Reconstituting a measure in a cruder form is not advisable unless there are strong reasons to do so, such as a clear multimodal score distribution.

The dummy variable approach can be extended to deal with categorical variables with more than two categories. Imagine that we trichotomize the professional relations variable, forming three approximately equal-sized groups that we label as "bad," "unsure," and "good." How can this variable be accommodated in a multiple regression using 0/1 dummy coding? A two-category variable can be represented with one dummy variable, as we have just seen. It can be shown that the information in a variable can always be represented with a *set* of dummy variables, when the size of the set is one less than the number of categories in the original variable. So for a three-category variable we need two dummy variables. Translating or coding a variable into a set of dummy variables can be accomplished in a variety of ways. One of the most commonly used is what is known as **reference coding.** In this coding system each dummy variable represents the mean difference on the dependent variable between a category and a fixed reference category chosen by the analyst. For example, we could choose the "bad" category as the reference group and create a dummy variable to represent the difference between "unsure" and "bad," and another to represent the difference between "good" and "bad."

If we rerun the multiple regression with the two dummy variables representing the trichotomized professional relations variable, the slope for the first difference is 5.6 and that for the second is 9.2, with only the latter being statistically significant with an alpha of .05. So using this form of measurement, it appears that professional relations does have an impact on mental health, but it is apparent only when the two extreme groups are contrasted. Note that this analysis does not include a comparison of the "unsure" and "good" categories, but this can easily be accomplished with another regression using a modified coding.

Another way of creating dummy variables is to use **ordinal coding.** Imagine that we retain our three categories but now want to treat them as ordinal: "good" is a *higher* category than "unsure," which is *higher* than "bad." Ordinal coding would again produce two dummy variables, but this time the first would represent the difference between "bad" and the other two categories, while the second would represent the difference between the "good" category and the other two. This is particularly helpful as, not only does it capture the ordinality of the measure, but it can also be used to detect threshold effects, that is, changes in mental health at particular points on the professional relations scale. Entering the ordinal dummy variables in the multiple regression provides a slope of 5.6 for the "bad"/other difference, and a slope of 3.6 for the "good"/other difference. It seems then that mental health differences are more marked for the "bad" group versus the rest. However, since neither slope is statistically significant in this analysis, the interpretation can be taken no further. (It should be noted, though, that using a technique we have yet to encounter, the two dummy variable are *jointly* significant. In other words, the ordinal measure *does* result in an association with mental health, but it cannot be broken down into specific comparisons within the measure in this instance.)

As the foregoing demonstrates, independent variables can be created in such a way that they can be included in regression analyses regardless of their measurement scale. These dummy variables may be used to deal with emergent measurement problems with variables that had initially been assumed to be interval in nature. Or, more important, they provide a way of combining truly categorical and ordinal variables with interval variables in a single regression analysis with full statistical control.

In addition to this flexibility, independent variables may be created in forms that help to check or combat violations of assumptions. As we noted in Section 4.3, for example, a skewed distribution may be corrected if the original scores are replaced with their square root. Or a nonlinear multivariate relationship may be made more linear by adding a "powered" independent variable. For example, we might wonder if the relationship between professional relations and mental health was curvilinear: Perhaps mental health benefits accrue only if a certain threshold level of professional relations is surpassed. This could be tested by running the standard multiple regression with not only the original four independent variables, but also a new variable in which professional relations scores are squared: a so-called quadratic term. Running this analysis shows that the slope for the squared variable is approximately zero and is highly nonsignificant. This indicates that the linear (unsquared) professional relations variable is sufficient to account for any relationship it may have with mental health. This should not be surprising as our bivariate and multivariate evaluation of the linearity assumption indicated no obvious violation.

Even after this cursory discussion, it should now be clear why ordinary least squares multiple regression is so popular. It provides within one framework a very flexible strategy for multivariate analysis of how multiple independent variables of all sorts impact on a dependent variable. It has a variety of troubleshooting tools and a range of solutions when data are ill behaved. Moreover, when these solutions are inadequate, other more complex forms of regression are available. But even within ordinary least squares regression there is yet another capability that we alluded to in Chapter 3 and can now explore more concretely: the use of series of regressions.

4.5 SEQUENTIAL REGRESSION ANALYSIS

The multiple regression strategy that we have concentrated on so far is called "all-in" because all of the independent variables are analyzed at the same time in one step. As we noted in Chapter 3, analytic power may be extended by conducting a series of regressions, each containing different subsets of the independent variables. Each subset may contain one variable or a block of variables. The subsets may be chosen by the analyst, in which case the strategy is called **hierarchical regression;** or they may be selected according to some statistical rule—an approach called **stepwise or statistical regression.** The primary focus of interest in sequential analyses is on how the picture *changes* from step to step. Does the arrival of additional independent variables produce an *increase* in explained variance? Does the beta for a particular independent variable *change* from one step to another as its companion variables change? This focus leads to a concern with *change* statistics, notably the change in R^2 from one step to another, and an F test of the null hypothesis that the R^2 *change* in the population is zero. These change statistics, in addition to the standard regression statistics we have already encountered, provide a powerful tool kit for analyzing complex patterns of relationships.

One such pattern is the moderating or interaction relationship whereby the *effect* of an independent variable on the dependent variable differs according to the *level* of a second independent variable. The use of hierarchical regression to examine a moderating effect can be exemplified with a recent study of goal conflict among German managers conducted by Kehr (2003). One of the many research questions addressed by this study was how changes in goal conflict over time impact on positive affect. Goal conflict and positive affect were measured on two occasions five months apart in a group of managers. These data were entered into a three-step hierarchical regression analysis with the later positive affect score as the dependent variable. At step 1 the earlier positive affect variable was entered as a control variable. This set a constant baseline against which the later affect score could be judged. At step 2 the earlier and later goal conflict variables were added. Then at step 3 an interaction variable was added that was literally the product of the two goal conflict variables. So by the final step there were four independent variables: the earlier positive le, the two goal conflict main effects, and the interaction es.

l particularly on the third step since it was the interaction effect suggested by the research question. It was argued goal conflict on concurrent positive affect differed accordearlier goal conflict, this would provide evidence that the the time pattern of goal conflict, and this proved to be the e earlier positive affect variable accounted for 16% of sitive affect ($p < .001$). The addition of the two goal conflict variables at step 2 produced no significant increase in R^2 and, consistent with this, neither of their betas was significant. Thus it appeared that goal conflict had no impact on positive affect. However, at step 3 the addition of the interaction variable produced a significant increase of 4% in R^2 ($p < .05$) and, equivalently, a significant beta ($p < .05$). In other words, goal conflict did have a concurrent impact on positive impact, but only when the earlier level of goal conflict was taken into account. Further analyses were conducted to uncover the pattern of this interaction. From these analyses it was determined that heightened goal conflict appeared to decrease concurrent positive affect, but only if earlier goal conflict had been low. In the author's words: "It was *emerging* rather than *enduring* conflict that was associated with a decrease in positive affect" (Kehr, 2003, p. 203, [original italics]).

Hierarchical regression can also be used to analyze mediating relationships whereby an independent variable impacts on a dependent variable through a second independent variable. In the simplest case, the middle variable in a chain of three mediates the impact of the first on the third variable. This type of analysis can be illustrated with an extract from an experimental study conducted by Emmons and McCullough (2003) that examined the effects of gratitude on positive affect. A group of college students was randomly assigned to one of two experimental conditions: a gratitude condition in which they were asked to record gratitude-inducing experiences over approximately a two-week period; or a hassles condition in which they were asked to record experiences that annoyed or bothered them. Each day they completed a number of measures including a gratitude feelings scale and a positive affect scale. Using these data, the authors addressed the question of whether the gratitude induction manipulation actually produced feelings of gratitude and subsequently increased positive affect. Thus they wished to know whether the monitoring of gratitude-inducing experiences would enhance positive affect and whether this effect would be mediated by the actual experience of grateful feelings.

Emmons and McCullough (2003) report their data analysis mainly in correlational terms, but the underlying strategy is essentially that of hierarchical regression and can be restated in those terms. The independent variable is the gratitude versus hassles manipulation that was treated as a dummy variable, the mediating variable is the gratitude feelings score averaged over time, and the dependent variable is the positive affect score similarly averaged. Before doing the mediation regression, it is necessary to be reassured that the three variables are linked at all. This was evident since the manipulation was correlated with gratitude feelings ($r = -.41$, $p < .001$), and gratitude feelings was correlated with positive affect ($r = .80$, $p < .001$). With this reassurance, a two-step hierarchical regression with positive affect as the dependent variable could be conducted to test the mediation effect as such. At the first step the manipulation variable produced a beta of $-.28$ ($p < .01$), showing that the gratitude induction procedure appeared to be associated with higher levels of positive affect. However, at the second step, when the gratitude feelings variable was added, the beta for the manipulation dropped to effectively zero (beta = .06, $p = .31$). In contrast, the beta of .85 ($p < .001$) for gratitude feelings showed that its relationship with positive affect remained strong. The change in the manipulation beta from step 1 to step 2 was interpreted as evidence for a total mediating effect since the association between the manipulation and positive affect disappeared when the gratitude feelings variable was removed from the chain by statistical control.

As we noted earlier, hierarchical regression requires the analyst to specify which subsets of independent variables will be used in a series of regressions. There is another sequential regression strategy in which no such specifications have to be made. Instead the selection of independent variables is done by a statistical program according to a set of rules. This may be accomplished by a so-called **forward selection** algorithm as follows. Imagine a data set with four independent variables and one dependent variable. A forward selection regression would first identify the independent variable that, *if* chosen at the first step, would produce the largest increase in R^2 and would be statistically significant. If a variable meets these requirements, it is selected and the procedure moves to the second step. Here another independent variable is chosen from the remaining three that would produce the largest increase in R^2 above that already achieved and that would be statistically significant. This second independent variable is selected, and the process continues until no more independent variables meet the significance criterion or the pool is empty. There are various ways to set the selection criteria, but they all revolve around the requirements for an independent variable to make the largest impact over and above variables that have already been chosen, and for this impact to be statistically significant.

The forward selection strategy builds up a set of independent variables from nothing. Alternatively, a reverse logic may be applied using the **backward elimination** strategy. The first step of this approach is an all-in regression with all independent variables selected. This time the search is for the independent variable whose elimination would produce the *smallest reduction* in R^2 and which was statistically *non*significant. This process continues until no more independent variables can be eliminated under the exclusion criteria. A third strategy—**stepwise regression**—is probably the most popular approach to statistical regression. The first step of this technique is the same as for the forward selection strategy: the selection of the independent variable that makes the biggest statistically significant impact on the dependent variable. But from the second step, the strategy becomes a mixture of forward selection and backward elimination. So at each step excluded variables are considered for inclusion and included variables are considered for exclusion. This process continues until all selection and elimination criteria are met or the pool of independent variables is exhausted.

The stepwise regression procedure can be exemplified with findings from a follow-up study of long-term mentally ill clients in Sweden conducted by Bjorkman and Hansson (2002). Their concern was to identify predictors of improvement in the quality of life of clients receiving case management over an 18-month period. In one of their analyses they used changes in global well-being as a dependent variable. The independent variables included in the analysis were changes in psychosocial functioning, global functioning, number of needs, social network, and psychiatric symptoms. A stepwise regression selected only two of these independent variables as significant predictors. At the first step social network change was selected as it accounted for 31.1% of the variance in global well-being change. At the second step change in psychiatric symptoms was selected as it explained a further 11.6% in the variance of the dependent variable. Both increases in R^2 were statistically significant at $p < .001$. The signs of the betas indicated that improvement in global well-being was associated with improving social networks and declining psychiatric symptoms. Once these two predictors were taken into account, no other independent variables displayed any statistically significant predictive capability.

Although we have discussed hierarchical and statistical regression separately, they can be combined to good effect. For example, the Bjorkman and Hansson (2002) analysis was actually a little more complicated than described in the last paragraph. They were concerned about the potentially confounding effects of clients' sex and age and of their level of self-esteem at the outset of the study. Accordingly, before the stepwise procedure was triggered, they forced these three variables into the analysis, which means that the effects of social network and psychiatric symptom changes on global well-being were over and above those due to sex, age, and self-esteem. Strictly speaking then, their analysis consisted of a one-step hierarchical regression followed by a two-step stepwise regression. This example also highlights the way in which independent variables may be treated in blocks rather than individually.

Of the regression techniques we have discussed, statistical regression attracts by far the most criticisms and cautions. Most of these focus on two issues: the problematic *p* values and the type of research question to which statistical regression provides answers. Earlier we noted the general problem of how multiple tests can inflate Type I error. It should be clear that this problem looms very large in statistical regression in which many "what if" regressions are carried out to select or eliminate independent variables. Moreover, these multiple analyses are on overlapping subsets of variables, so it becomes technically difficult to estimate exact *p* values. This is worrying because the whole process typically depends heavily on *p* values as criteria for including or excluding variables. It is possible to use adjusted values (Tabachnick & Fidell, 2001, pp. 142–143), but they do not provide a complete solution to this problem.

The second issue is more conceptual than statistical and may be framed in the following question: What does statistical regression tell you that you actually want to know? The Swedish study can be used to show how results might be misinterpreted, though without attributing these misinterpretations to the authors. It is tempting to conclude from this analysis described above that changes in social network and symptoms were the *best* predictors of changes in global well-being, but this would be erroneous, or at least unwarranted. To find the best predictors, it would be necessary to search among all possible subsets of the independent variables, and stepwise regression does not do this. It reviews many possibilities but far from all. In fact, the search process is highly constrained by the selection criteria that focus on ordering by magnitude and by net gain. So the social network variable was chosen first because it was the best single predictor. The symptom variable was chosen at the second step because it made the next biggest contribution, net of the social network contribution. This is an elegant statistical process, but the question is how well it reflects the logic and requirements of a particular research question. It may well fit in a highly applied research context in which some guide to action is needed that maximizes efficiency and cost benefits. But the ordered, net gain logic does not seem to apply to many research questions, and often satisfactory answers can be found with a straightforward standard regression.

4.6 FURTHER READING

Darlington's (1990) text provides an excellent, accessible introduction to multiple regression. More detailed and technical discussions can be found in the classic texts of Cohen et al. (2003) and Pedhazur (1997). The former is a particularly good source of information on the more esoteric forms of regression alluded to in Subsection 4.3.2 in the present chapter. Tabachnick and Fidell (2001, Chapter 5), Hair et al. (1998, Chapter 4), and Stevens (2002, Chapter 3) all give helpful accounts that focus more on computer analyses. Achen (1982) provides a brief but very thought-provoking discussion on the interpretation of regression statistics. On a more specific topic, Jaccard and Turrisi (2003) have produced a particularly helpful guide to testing moderating effects with multiple regression.

FIVE

LOGISTIC REGRESSION AND DISCRIMINANT ANALYSIS

In the previous chapter, multiple regression was presented as a flexible technique for analyzing the relationships between multiple independent variables and a single dependent variable. Much of its flexibility is due to the way in which all sorts of independent variables can be accommodated. However, this flexibility stops short of allowing a *dependent* variable consisting of categories. How then can the analyst deal with data representing multiple independent variables and a categorical dependent variable? How can independent variables be used to account for differences in categories?

This chapter introduces two techniques for accomplishing this aim: logistic regression and discriminant analysis. Even though the two techniques often reveal the same patterns in a set of data, they do so in different ways and require different assumptions. As the name implies, logistic regression draws on much of the same logic as ordinary least squares regression, so it is helpful to discuss it first, immediately after Chapter 4. Discriminant analysis sits alongside multivariate analysis of variance, the topic of Chapter 6, so discussing it second will help to build a bridge across the present chapter and the next. That said, the multivariate strategy of forming a composite of weighted independent variables remains central, despite differences in the ways in which it is accomplished.

In Subsection 5.1.1 we explore the nature of the weighted composite variable in logistic regression with a dichotomous dependent variable and introduce the main statistical tools that accompany it. Subsection 5.1.2 shows two-group, or "binary," logistic regression in action, first with further analyses of the nurses' data introduced in Chapter 4 and then with examples from the research literature. In Subsection 5.1.3 the usual questions of trustworthiness will be raised with specific reference to logistic regression. Then in Subsection 5.1.4 extensions to the basic technique are discussed, including how to deal with different types of independent variables and with a dependent variable that has more than two categories. Subsections 5.1.3 and 5.1.4 will be relatively brief since they will draw heavily on material we have already covered in Chapter 4 on multiple regression. The second half of this chapter, comprising Subsections 5.2.1–5.2.4, follows the same sequence of topics for discriminant analysis.

5.1 LOGISTIC REGRESSION

5.1.1 The Composite Variable in Logistic Regression

Although it is inappropriate to use ordinary least squares (OLS) regression when the dependent variable is categorical, it is instructive to begin by asking how the composite variable would function if OLS regression were used. In its most general form the relationship between multiple independent variables (IVs) and a single dependent variable (DV) is:

DV = [coefficient 1(effect 1) + coefficient 2(effect 2) + ...
+ constant]
+ residual

For OLS regression, this general expression becomes:

DV = [slope 1(IV1) + slope 2(IV2) + ...
+ Y intercept]
+ residual

The composite variable in the square brackets generates predicted scores on the dependent or Y variable. Values for the slopes and Y intercept are chosen that maximize the correlation between the actual and predicted Y scores or, equivalently, minimize the gap or residual between them.

What happens if this strategy is applied to data in which the dependent variable consists of two categories, labeled 0 and 1 (i.e., a dummy variable)?

The composite cannot be used to generate predicted *scores* on the dependent variable since there are no scores to predict, only categories. Instead the composite now generates the predicted *probability* of a case being in the category labeled 1. These predicted probability values should lie between 0 and 1 and can be subtracted from the actual 0 and 1 values to obtain residuals. The regression slope will have the usual interpretation, except that it will be in probability terms: for every 1-unit change in a given independent variable there will be a change in probability of being in category 1, which is equivalent to the slope value. All of this makes it sound as if OLS regression is well suited to a categorical dependent variable, so where is the problem?

Actually, there are several problems that have been detailed with great clarity by Pampel (2000), to whose primer on logistic regression the present account is much indebted. In summary, using OLS regression to generate predicted probabilities can produce values outside the 0 to 1 range, forces linearity on what is more likely an S-shaped relationship, violates the assumption that the components of the composite variable are additive, and violates the assumptions of normality and homoscedasticity required for statistical tests. After such a list of charges, there seems little option but to seek an alternative strategy. The logistic regression strategy retains the goal of generating predicted probabilities but achieves it indirectly by using another probability index and a different criterion to choose the coefficients in the composite variable. These two "moves" make for a convoluted and abstract journey from the data to the results. We will just highlight the landmarks along the way and as usual emphasize the familiarity of the big road map.

In the present context the probability of being in one of two groups is provided by the relative frequency, that is, the number of cases in one group divided by the number of cases in both groups. If group 1 contained 80 cases and group 0 contained 20 cases, the probability of being in group 1 would be 80/100 = .8 or 80%. This is the type of probability that we are trying to predict, but that is inadequately predicted using OLS regression. To obtain more accurate predicted probabilities, the first step is to focus on another type of probability index that we encountered in Chapter 1: the odds. The odds of being in group 1 for our imaginary 100 cases would be 80/20 = 4. A case is four times more likely to be in group 1 than in group 0. Since the probability and the odds combine the same frequencies in different ways, they are obviously closely related (the probability is just the odds divided by the odds plus 1). But this simple move opens the door to a solution to the problem of predicting probabilities.

The next step within the first "move" is to change the scale of the odds by transforming it, for reasons that will become apparent shortly. The specific transformation is to replace the odds with its natural log. The natural log of a number is the power to which 2.718 has to be raised to produce that number. So now, instead of dealing with odds, we are confronting log odds, also known as **logits.** What is the payoff for this mind-numbing shift into mathematical abstraction? It can be summed up in the following:

predicted log odds of a DV = [logistic coefficient 1(IV1)
+ logistic coefficient 2(IV2)
+ . . . . + constant]

Working with log odds rather than probabilities as such means that the familiar composite of independent variables is applicable and retains all its usual properties. In terms of the problems raised earlier, the composite will capture an S-shaped relationship between the independent and dependent variables, it will be additive, and the predicted probabilities that can be derived from it will fall between 0 and 1. The logistic coefficients will be interpretable as statistically controlled effects as usual although, since they are on a log odds scale, they will require some massaging to be useful. But before we delve into this sort of detail, we need to ask how the logistic coefficients including the constant are obtained: the second move in the overall strategy.

As we just noted, a predicted probability for each case can be derived from the log odds and consequently so can a residual—the difference between the prediction for that case and their actual 1 or 0 status. However, the regression coefficients that minimize the residuals' sum of squares for all the cases, that is, that meet the ordinary least squares criterion, will not necessarily maximize predictive power. Moreover, any statistical tests that are based on this way of choosing coefficients will violate the assumptions of normality and homoscedasticity and produce inaccurate p values. To avoid these problems, a different criterion for selecting coefficients is adopted: the criterion of **maximum likelihood.**

Under this criterion, the aim is still to minimize the difference between a case's predicted probability of being in a category and its actual category. The search is for the coefficients that will produce the log odds that in turn produce the predicted probabilities that will most accurately place cases in their actual category. So the maximum likelihood criterion produces the logistic coefficients that will most closely reproduce the actual categories in which cases appear. The predicted probabilities and actual categories for each case are bundled up, not into a sum of squares package for all cases, but into a statistic called the **log likelihood function.** So, in an opaque nutshell, the aim is to find the coefficients that maximize the value of the log likelihood function. To make matters even more opaque, the log likelihood function is often multiplied by –2 to turn it into the **log likelihood chi²** statistic, as we will see shortly. Multiplying by –2 also means that the log likelihood values range from 0 to positive infinity and that the strategic aim is now to find coefficients that *minimize* the value of this function. These rapid turnabouts should become less dizzying when we look at logistic regression in action, below.

To summarize, the relationships between multiple independent variables and a categorical dependent variable can be analyzed using a technique called logistic regression. This involves forming the independent variables into the usual weighted, additive composite, which is then used to predict the probability of cases appearing in a particular category of the dependent variable. However, in order to achieve this legitimately, two moves are made. First, the predicted probabilities are derived indirectly through logged odds, or logits. Second, the coefficients in the composite are calculated using a procedure called maximum likelihood estimation. A set of coefficients is chosen provisionally that, through log odds, generates the probability of each case being in a given category. These probabilities and the actual category memberships are fed into the log likelihood function, which produces a particular log likelihood value. Then different sets of coefficients are tried and those that produce the maximum log likelihood value are the ones that are finally selected as the logistic coefficients. To make all of this more concrete, and to see what other statistics result, we now turn to some actual logistic regression analyses.

5.1.2 Binary Logistic Regression in Action

A large part of Chapter 4 was spent in exploring the use of multiple regression to analyze the effects of workplace characteristics on the mental health of a group of nurses (Budge, Carryer & Wood, 2003). To highlight the similarities and differences between ordinary least squares and logistic regression, we will now return to this data set. We will reanalyze the relationships between workplace characteristics and mental health, but with the latter now treated as a dichotomous categorical variable. So for present purposes, the data

Table 5.1 Mean Scores and F Tests on Three Workplace Characteristics for Nurses in Good and Poor Mental Health

Variable	*Good Health Mean*	*Poor Health Mean*	*F*	*p*
Professional relations	14.70	13.40	8.35	.004
Autonomy	13.60	13.10	1.23	.269
Control	17.90	17.20	0.82	.368

set consists of three independent variables: professional relations, autonomy, and control; and a dichotomous dependent variable: mental health coded 1 for good health and 0 for poor health. Of the 163 nurses in the sample, 91 reported good mental health and 72 reported poor health. The earlier multiple regression analyses also included age as an independent variable, but since it had no effect on mental health and adds distracting detail, it has been omitted from the present analyses.

Before we embark on the logistic regression, it is helpful to gain a bivariate overview of the data, just as an inspection of bivariate correlations is advisable in multiple regression. Since we are interested in the relationships between interval and categorical variables, we can make use of mean differences and ANOVAs (or t tests) to achieve this. Table 5.1 summarizes the bivariate relationships between the independent and dependent variables.

The mean differences indicate that nurses in good mental health report better relations, better autonomy, and better control than those in poor health. However, the F tests suggest that only in the case of professional relations is the difference statistically significant. These separate bivariate analyses are informative, but they fail to take into account the correlations among the three work variables. Since professional relations correlates .56 with autonomy and .48 with control, and autonomy correlates .69 with control, it is quite possible that the conclusions we have just drawn are distorted by confounding. This is one of the fundamental reasons for turning to a multivariate analysis, in this case logistic regression analysis. As usual, we take a top-down approach, beginning with the performance of the composite variable overall and then proceeding to examine particular coefficients, if this appears justified.

The −2 log likelihood statistic of 215.15, which is the lowest value that emerged from trying out different sets of coefficients, reflects the multivariate relationship. Since it is significant by chi^2 test at $p = .035$, we can conclude that there is a statistically significant relationship between the set of independent variables and the dependent variable, if we adopt an alpha of .05. But what exactly is the null hypothesis under test here? Testing omnibus hypotheses in logistic regression is an inherently comparative exercise. In fact, this is usually the case in statistical analysis, but here it becomes more explicit. As we have noted, a set of independent variables or effects is referred to as a "model," so more precisely we are engaged in a comparison of models. In the present case the model containing the three independent variables is being compared with a model that contains only the constant. In other words, we are testing whether knowledge of the workplace characteristics improves our ability to predict mental health status. If we were to do a logistic regression *without* the independent variables, the "baseline" –2 log likelihood would be 223.75. Including the independent variables reduces the –2 log likelihood to 215.15: an improvement of 8.6. It is this *change* that is indexed and tested by the log likelihood chi^2 statistic, the null hypothesis being that there is no change in the population. This is conceptually parallel to the statistical testing of R^2 change in sequential regression. At the first step of a stepwise multiple regression, for example, the statistical test can be seen as being of the *change* in R^2 from zero, when there are no independent variables present, to whatever value it achieves when the first independent variable enters.

This reference to R^2 in multiple regression highlights the fact that, in logistic regression, there is no straightforward index of the strength of the multivariate relationship between the independent and dependent variables. This should not be surprising given the grounding of R^2 in sums of squares, which are now notable by their absence. The log likelihood statistics are useful for hypothesis testing but do not offer an interpretable measure of association. Various attempts have been made to develop "pseudo" R^2 statistics for logistic regression. For example, the SPSS program provides the Cox & Snell and the Nagelkerke pseudo R^2 statistics, which are .051 and .069, respectively, in the present analysis. So we could tentatively conclude that the three independent variables explain between 5.1% and 6.9% of the variance in mental health. However, there are a variety of such pseudo statistics, all giving different estimates, and none regarded as superior to all the others, so they are best treated with caution if not actually avoided. Note this means that, aside from these pseudo statistics, logistic regression statistics therefore inherently focus on group differences rather than individual differences.

There is another method of expressing the strength of the multivariate relationship, which is not only less contentious but also more intuitively appealing and potentially more practicable. This method uses the predicted probabilities to assign cases into the categories of the dependent variable and then compares the results with their actual categories. Cross-classifying cases according to their assigned and actual categories provides another picture of how well the independent variables predict the dependent variable. Since the predicted probabilities are decimal values between 0 and 1, they need to be dichotomized so that they can be compared with the actual 0 and 1 categories in a 2 × 2 table. In the present analysis, cases with a predicted probability below .5 were assigned to the 0 category, and those with a value above .5 were assigned to the 1 category.

In the nurses' sample about 82% of those in good mental health were correctly classified, while only 33% of those in poor health were accurately predicted. This gives an overall "hit rate" of about 61%. This sounds impressive, especially in the good health group, but these figures have to be compared with what could be achieved even in the absence of any knowledge about the nurses' workplaces. In such a situation one prediction strategy would be to assign all cases to the modal category, the one that actually contains most of the cases. This is the good health category that contains 91 of the 163 cases. Following this prediction rule would result in a 100% hit rate for the good health category and 0% for the poor health category. Overall this would give an average hit rate of about 56%. So, on this strategy, using the workplace predictors increases the overall hit rate from 56% to 61%: a gain of only 5%, but at the cost of a disastrous hit rate for the poor health group.

A less draconian strategy would simply be to use the relative frequencies in the sample as a basis for assignment: 56% for the good health group and 44% for the bad health group. If this defined the baseline or chance expectation, the gain in hit rate for the good health group would be 26% (82% – 56%) but *minus* 11% for the poor health group. Details aside, the key points to note in classification analysis are that hit rates need to be compared with chance and that chance can be interpreted in more than one way. We will return to this issue in Subsection 5.2.2 as classification analysis is also used as part of discriminant analysis.

At this point the results indicate that the set of work characteristics is related to mental health to a degree that is unlikely to be due to chance. The magnitude of this relationship is hard to specify precisely, but the pseudo R^2 statistics in the range 5%–7%, and the gain in predictive hit rate, suggest that the relationship is probably weak. Given that there is some relationship, which independent variables are contributing to it? The answer to this question can be found in the logistic coefficients and their associated statistical tests, which may be z tests or the Wald tests shown in Table 5.2.

Table 5.2 Logistic Regression Statistics Showing the Effects of Three Workplace Characteristics on Mental Health

Independent Variable	*Coefficient*	*Wald Statistic*	*p*	*Odds*
Professional relations	.193	6.753	.009	1.212
Autonomy	−.027	.116	.734	.973
Control	−.012	.057	.812	.988
Constant	−1.907			

The SPSS program uses the Wald statistic to test the null hypothesis that a coefficient value is zero in the population. As Table 5.2 shows, only in the case of the professional relations variable should this null hypothesis be rejected with an associated p value of .009. Under some circumstances the Wald statistic can produce misleading results, and so it is wise to check the pattern of results by comparing models. In the present situation the question of interest would be whether the model containing professional relations and the constant would perform better than the constant-only model. In other words, does the professional relations variable truly have predictive power in the absence of the other two predictors?

As we saw earlier, the −2 log likelihood for the constant-only model is 223.75. Running a logistic regression that includes professional relations reduces this figure to 215.55, a reduction and chi^2 of 8.2 that is statistically significant with a p value of .004. Moreover, this professional relations model has pseudo R^2 statistics in the range 4.9%–6.6% and a similar gain in hit rate. The significant chi^2 test for the professional relations model, and the similarity of the magnitude statistics in the models containing one or three independent variables, strongly suggest that only the professional relations variable is contributing to differences in mental health.

The logistic coefficient for professional relations of .193 indicates its impact on mental health when the other two independent variables are statistically controlled. However, the fact that it is on a log odds scale means that it is not easy to interpret. The coefficient says that nurses who are higher by one unit on the professional relations scale have a .193 increase in their log odds of being in the good health group. As usual, the coefficients can be positive, as in this case, or negative if the relationship is inverse. The coefficients can be turned into probabilities, but these are even more difficult to interpret because the impact is not uniform across the independent variable scale. Instead the most common strategy is to convert the coefficients into odds, and these appear in the last column of Table 5.2. Remember that an odds of 1 indicates no relationship, a value greater than 1 indicates a positive relationship, and a value less than 1 indicates a negative relationship. The odds statistics can be interpreted in terms of percentage change by subtracting 1 and multiplying by 100. So the odds of 1.212 for professional relations mean that for every 1-unit increase in that independent variable, the odds of being in the good mental health group increase by 21.2%. Every unit increase in autonomy produces a 2.7% *decrease* in the odds [100(.973 – 1) = –2.7] and for control the decrease in odds is 1.2%. Bear in mind, though, that the coefficients for autonomy and control are not statistically significant, so we should be treating them as effectively zero and their corresponding odds as 1. The description here is purely for illustrative purposes.

As in OLS regression, confidence intervals may be calculated around logistic coefficients and around the odds. For example, the 95% confidence interval for the professional relations odds ranges from 1.048 to 1.402. So, while the best single odds estimate is 1.212, we can be 95% confident that the population value lies within this range. Finally, it is important to appreciate that the logistic coefficients are unstandardized, and therefore not directly comparable with each other unless the independent variables happen to share the same unit of measurement. According to Pampel (2000), while there are various ways to calculate standardized coefficients, none are truly equivalent to the betas found in ordinary least squares regression. A partial solution is to standardize the independent variables, either before they are entered into the analysis or by multiplying the coefficient for a variable by its standard deviation. However, since the dependent variable remains in its original form, this semistandardization is only a semisolution.

Now that we have discussed the most commonly used statistics in logistic regression, two examples from the research literature may help to consolidate understanding. Kirschenbaum, Oigenblick, and Goldberg (2000) used binary logistic regression to examine differences between two groups of Israeli workers: 77 who had suffered a first-time work injury and 123 who had suffered injuries on multiple occasions. The independent variables were a variety of sociodemographic, work environment, and well-being indicators. For a logistic regression containing 11 independent variables, they report a chi^2 of 15.9 with a p value of $< .01$. Clearly, there was a multivariate relationship between the independent variables and the work injury categories that was unlikely to be due to chance. A classification analysis showed that 140 of the 200 cases were correctly classified by the probabilities derived from the model: an overall hit rate of 70%.

Turning to the independent variables, we find that 11 of the 17 logistic coefficients were statistically significant at $p < .05$. Three of these were well-being variables: a feeling that things are going wrong, unhappy with family life, and unhappy with housing. Thus it appears that some aspects of well-being had an influence on proneness to multiple work injuries. However, inspection of the *signs* of the coefficients revealed an anomalous pattern. Multiple injuries were more likely for workers who were unhappy with their housing (coefficient = 3.396), but *less* likely for those who felt that things were going wrong (coefficient = −1.657) or who were unhappy with family life (coefficient = −2.66). The authors then provide some interesting suggestions on how this apparent anomaly might be resolved.

Many reports of logistic regression analyses omit information about the chi^2 for the model, classification results, and the logistic coefficients. Instead they focus on the odds for each independent variable, often including the 95% confidence interval rather than p values. Natvig, Albrektsen, and Qvamstrom (2003), for example, analyzed the predictors of happiness in a sample of 887 Norwegian school adolescents. In one logistic regression the dichotomous dependent variable was very or quite happy versus not happy, and the independent variables were school alienation, school distress, general self-efficacy, school self-efficacy, support from teacher, support from pupils, and decision control. The logistic regression results showed that several of these variables had odds with a 95% confidence interval that did not include 1. These are the variables whose effects would be statistically significant if null hypothesis testing were used with an alpha of .05. The school alienation and general self-efficacy variables exemplify this and can be used to reiterate how odds statistics are interpreted.

The school alienation odds of .47 were less than 1, indicating a negative relationship with happiness. The dependent variable was coded such that the odds are those of being in the very or quite happy group. Subtracting 1 from .47 and multiplying by 100 indicates that with every 1-unit increase in school alienation the odds of being happy decreased by 53%. The confidence interval revealed that in the population this decrease could be as great as 69% or as little as 27%. Turning to general self-efficacy, we find that the odds of 1.7 mean that for every 1-unit increase in that variable, the odds of being happy increased by 70%. Again, the confidence interval suggests that we can be 95% confident that the population value for this increase lies between 10% and 163%. Since this is a multivariate logistic analysis, the odds have been adjusted to take account of any associations among the independent variables and, in this example, they have also been adjusted to control for age, gender, and school.

5.1.3 Trustworthiness in Logistic Regression

The issues that bear on the trustworthiness of logistic regression results can be discussed briefly since the details of most of them have already been explored with respect to OLS regression in Section 4.3 of Chapter 4 and more generally in Chapter 2. We will follow the usual sequence of first considering sampling and measurement issues, then the assumptions required for the legitimate use of the technique, and finish with some other general concerns.

The sample size required for logistic regression is typically greater than that needed for OLS regression. Statistical tests of coefficients obtained by maximum likelihood estimation may give misleading results for samples under 100 (Pampel, 2000, p. 30). More independent variables require more cases, and a minimum of 50 cases per independent variable is recommended (Wright, 1995, p. 221). As usual, the appropriate sample size for a given analysis is also dependent on the acceptable levels of Type I and II error, the expected magnitude of the relationships between the independent and dependent variables, the reliability of measurement, and the frequency distribution of the dependent variable. In the logistic regression context, the more unequal the numbers in the categories, the more cases are needed. Add to all this the problem of missing data because of listwise deletion, and the desirability of having enough cases to cross-validate results on a holdout sample, and it becomes painfully clear that logistic regression typically requires cases in the hundreds to guarantee trustworthy results.

Regarding measurement, the independent variables may be on any type of scale, and they are dealt with as in OLS regression, using dummy coding where necessary. The dependent variable is usually categorical and may have two or more categories, as we will see in the next section. Also in the next section, it will become apparent that the dependent variable may be on an ordinal scale, again by using dummy coding. However the dependent variable is scaled, it is required that the categories are mutually exclusive and jointly comprehensive. So each case must be locatable in one and only one category. As usual, it is assumed that the data have been produced by reliable and valid measurement procedures. It is important to emphasize that assigning cases to categories may be a highly complex and error-prone process. A simple outcome of a measurement procedure does not mean that the procedure itself is simple.

The assumptions required for statistical tests in logistic regression are far less restrictive than those for OLS regression. There is no formal requirement for multivariate normality, homoscedasticity, or linearity of the independent variables within each category of the dependent variable. However, as Tabachnick and Fidell (2001, p. 521) note, satisfying these conditions among the independent variables for the whole sample may enhance power. The problem of multicollinearity—very high correlations among the independent variables—does apply to logistic regression. All of these assumptions about the independent variables may be evaluated by treating one of the independent variables as a pseudodependent variable and regressing it on all the other independent variables using OLS regression. The tenability of the assumptions can then be examined with the usual OLS diagnostic tools. The assumption of independence of cases remains in place. In other words, each case can appear in the data set only once, and their data must be uncorrelated with the data of any other case. Casewise exploration of the residuals—the difference between the predicted probability and the actual category—may reveal patterns suggesting nonindependence and may identify outliers for whom the model provides notably poor predictions.

To this point we have concentrated on trustworthiness issues that arise in one form or another in regression generally. Two further issues present themselves in logistic but not in OLS regression. The first issue is that the maximum likelihood procedure for estimating the logistic coefficients is an **iterative procedure.** This means that the coefficient values are calculated in a series of steps or iterations rather than in one hit, as in OLS regression. The aim at each iteration is to produce a log likelihood that is greater than that at the preceding iteration. This process continues until a convergence criterion is satisfied, that is, the amount of increase between two iterations is small enough for the solution to be regarded as stable. In some circumstances the procedure may fail to converge on estimates of the coefficients, either because the convergence criterion could not be met or the number of permitted iterations was exceeded. Or less dramatically, convergence may be achieved, but only at the cost of a large number of iterations. In all of these situations a warning signal is being given that the data are problematic in some way, and the results may not be trustworthy. So information about the iteration history of a maximum likelihood analysis can be another useful diagnostic tool.

The second issue concerns the trustworthiness of classification analyses. Classification results are never definitive because they depend on at least two decisions made by the analyst that may be questionable. The first is the cutpoint used to translate predicted probabilities into predicted categories. The usual default is .5, but this may not be the optimum choice. The second decision concerns the best choice of baseline hit rate against which the achieved hit rates should be judged: an issue we noted earlier. This could be the simple probability (50% in a two category analysis), the relative frequency in the sample, the relative frequency taken from available population data, or a figure based on some other criterion. A different choice of baseline hit rate can give a very different sense of the predictive power of a given model. Even when appropriate decisions on these two issues have been made, the concreteness of classification analysis can also distract from the point that the hit rates are generated from and tested on the same data. This capitalization on chance means that hit rates in any replication are almost inevitably going to be less impressive. Accordingly, when possible it is advisable to generate predictions with one subgroup from the sample and to test their predictive power on another "holdout" subgroup. Failing this, other cross-validation techniques can be used *within* one sample to test the stability of the results across different subsets of the sample.

5.1.4 Extending the Scope of Logistic Regression

Like OLS regression, logistic regression can accommodate independent variables on any measurement scale with the use of dummy coding. For example, Hintikka (2001) examined the relationship between religious attendance and life satisfaction in a random sample of 1,642 adults in Finland, using almost entirely categorical variables. Both of these variables had two categories, while the control variables of sex, employment status, household category, and adequate social support had two, three, or four categories. Age was the only independent variable that was not categorical. A binary logistic regression was conducted to assess the relationship between religious attendance and life satisfaction, while controlling for all of the other independent variables. Hintikka reports an adjusted odds for religious attendance of 1.7 with a 95% confidence interval of 1.2–2.4. This means that religious attenders were 70% more likely than nonattenders to be satisfied with their lives. Alternatively, we could say that religious attenders were 1.7 times more likely than nonattenders to be satisfied with their lives.

Logistic regression can also be used to evaluate interaction or moderating effects, using the products of independent variables, as in OLS regression. The study by Natvig et al. (2003) of happiness in school adolescents, discussed earlier, included such interaction terms. Thus their logistic model included not only the independent variables described earlier, but also the product of each separately with age and sex. Since none of these interaction variables were statistically significant, it could be concluded that the predictors of happiness that were found were not moderated by age or sex.

All of the sequential techniques used in OLS regression are also available in logistic regression. Kirschenbaum et al.'s (2000) analysis of the predictors of work accident proneness was actually more complex than the description given earlier. As noted then, the independent variables fell into three groups: sociodemographic, work environment, and well-being characteristics. The analytic strategy was to build a hierarchical model where these blocks of variables were entered in three cumulative steps. This allowed the analysts to examine the predictive gain at each step and to note changes in the coefficient for a particular variable at each step. For example, the coefficient for sex changed from a statistically significant 1.201 at step 1 to a nonsignificant .742 at step 2 and then dropped further to .418 at step 3. Such a pattern suggests confounding or possibly mediation if other conditions were met. In fact, even this description understates the complexity of the analysis because the selection of particular variables into the blocks was guided by earlier forward and backward logistic regressions. That is, variables were selected for inclusion in the blocks according to statistical rules rather than by the analysts.

The final extension of logistic regression in this section concerns the structure of the dependent variable. To this point we have focused on binary logistic regression, which allows for a two-category dependent variable. More than two categories can be accommodated with the technique of **multinomial or polytomous logistic regression.** To achieve this, the categories are converted into a set of dummy variables, one less than the number of categories.

Each dummy variable represents a particular difference between particular categories, either singly or in sets. As we saw in Chapter 4, two systems of particular interest are reference and ordinal coding. In the first, a particular category is chosen as a reference, and each dummy variable represents a difference between that category and each of the others. In ordinal coding the ordinal scaling of the dependent variable is represented by a set of dummy variables, each representing a comparison of the sets of categories above and below each scale point.

In multinomial logistic regression, a logistic model is estimated for each dummy dependent variable. This means that no new interpretive issues arise, and the only concern is to be clear about the particular difference that a given model is estimating. Returning to the study of the predictors of happiness in adolescent schoolchildren (Natvig et al., 2003), the researchers conducted both binary and multinomial logistic regressions. In the former, as we saw earlier, the dependent variable was very or quite happy versus not happy. For the multinomial analysis, they created two dummy variables: very happy versus not happy, and quite happy versus not happy, to represent three categories of happiness. This is an example of reference dummy coding with not happy as the reference category.

Earlier in the binary logistic regression we saw that one of the successful predictors, school alienation, differentiated between the very or quite happy and not happy categories with an odds of .47. The multinomial regression produced odds of .53 and .35 for the very happy versus not happy and quite happy versus not happy contrasts, respectively, and neither of their 95% confidence intervals included 1. This means that school alienation differentiates the not happy category from the other two, both singly and jointly, and this pattern is unlikely to be due to chance. However, a different pattern emerged for the other successful predictor—general self-efficacy. For this variable, the binary odds for the very or quite happy versus not happy comparison were 1.7. The multinomial regression produced odds of 1.39 and 2.89 for the very happy versus not happy and quite happy versus not happy contrasts, respectively, but the 95% confidence interval for the former did not include 1. Accordingly, it appears that general self-efficacy did not predict the difference between the quite happy and not happy categories. Details aside, it should be apparent that multinomial logistic regression provides the capacity not only to accommodate a variety of categorical and ordinal dependent variables, but also to detect specific differences between categories within these variables.

This completes our introduction to logistic regression. We now turn to an alternative multivariate technique for analyzing the relationships between multiple independent variables and a single categorical variable: discriminant analysis.

5.2 DISCRIMINANT ANALYSIS

5.2.1 The Composite Variable in Discriminant Analysis

Discriminant analysis captures the relationship between multiple independent variables and a categorical dependent variable in the usual multivariate way, by forming a composite of the independent variables. So, discriminant analysis and logistic regression can be used to address the same types of research question. As in logistic regression, the variable generated by the composite cannot be a predicted score on the dependent variable. Instead it is a **discriminant function score** that then feeds into calculations that produce the predicted probability of a case being in a particular category of the dependent variable. This predicted probability is then used to generate a predicted category for each case. So, in broad terms the strategy is very similar to logistic regression in which the composite variable generates logits, which produce predicted probabilities, which produce predicted categories. The composite variable in two-group discriminant analysis is:

discriminant score = [discriminant coefficient 1(IV1)
+ discriminant coefficient 2(IV2) . . . .
+ constant]

The coefficients are now called discriminant function coefficients. For each case, the coefficient for an independent variable is multiplied by the case's score on that variable; these products are summed and added to the constant; and the result is a composite score for that case—their discriminant score. From these scores can be derived predicted probabilities and predicted group membership on the dependent variable.

Before we look more closely at the coefficients, it would be helpful to discuss the principle by which they are calculated. This principle will be clearer if we first pause to appreciate the hybrid nature of discriminant analysis and to review briefly some material from Chapter 1. When we consider the typical interpretation and application of the technique, it is convenient to frame it in regression terms: the prediction of a categorical dependent variable using multiple independent variables. However, it is easier to appreciate *how* the technique works if we frame it as a form of multivariate analysis of variance (MANOVA), which indeed it is. MANOVA is the multivariate form of ANOVA in which there are multiple *dependent* variables. (This is discussed in detail in Chapter 6.) The unsettling consequence of this shift in perspective is that we need to reverse the status of the independent and dependent variables temporarily. So from a MANOVA perspective we are now asking how well a categorical variable accounts for differences in a set of *dependent variables.* To make this more concrete, in the next section we will return to the nurses' data and the relationship that autonomy, control, and professional relations have with good versus poor mental health. From a regression perspective, there are three independent variables and one dependent variable. But from the MANOVA perspective that we now adopt temporarily, we have one independent categorical variable and three dependent variables. This may sound like cheating, but in fact it just highlights the way in which independent and dependent variable status is something imposed by the analyst rather than embedded in the statistics.

Imagine that we want to analyze the bivariate relationship between the nurses' mental health, treated as a dichotomous independent variable, and their professional relations treated as a dependent variable. In Chapter 1 we saw how analysis of variance can be used to analyze the relationship between a categorical independent variable and an interval-level dependent variable. In fact, we quickly carried out an ANOVA on the relationship between mental health and professional relations in Subsection 5.1.2 of this chapter. At the heart of the ANOVA strategy is the idea of capturing group differences on the dependent variable with a between-groups sum of squares and individual differences with a within-group sum of squares. The between-group and within-group sums of squares add up to the total sum of squares, which represents all of the individual differences on the dependent variable, regardless of group. The basic rationale of this approach is that the bigger the between-groups sum of squares is relative to the within-group sum of squares, the more likely it is that the independent and dependent variables are related. In Chapter 1 we saw how this relationship can be indexed with the ratio of between-group/total sum of squares (eta^2 = explained variability), or of within-group/total sum of squares (Wilks's lambda = unexplained variability). Further, the ratio of between-group/within-group sum of squares can be changed into a ratio of variances that then becomes the test statistic known as the F ratio.

For the nurses' data, those in good mental health have a mean professional relations score of 14.7, while those in poor mental health have a mean score of 13.4: a mean difference of 1.3, as we saw in Table 5.1. Analysis of variance gives an eta^2 of .049, a Wilks's lambda of .951, and an F of 8.35 ($p = .004$). The F statistic reassures that the relationship is unlikely to be due to chance, and the other statistics indicate that mental health status accounts for 4.9% of variance in professional relations or, conversely, that it leaves 95.1% of variance unexplained. As we move into discriminant analysis, we carry forward from this bivariate analysis two particular perspectives. The first focuses on the *distance* between the two means. The second focuses on the ratio of the between-groups/within-groups sum of squares, which lies at the heart of the F ratio. This sum of squares ratio is known as the **eigenvalue,** and it is the statistic on which discriminant analysis pivots.

In discriminant analysis the ANOVA logic we have been reviewing is applied to the composite variable: the discriminant score. If we return to the example in which there are three quasi-dependent variables (autonomy, control, and professional relations), each nurse will have a discriminant score that combines their weighted scores on these three variables. These discriminant scores can be divided into the good and poor mental health groups, and the mean discriminant score can be calculated for each group. The group means on the composite variable are known as **centroids.** Now we are finally in a position to state the principle by which the discriminant coefficients or weights are selected. They are chosen so that the distance between the centroids is maximized, within certain constraints that need not concern us. So coefficients are chosen that push the group means on the composite variable as far apart as possible, that is, that maximally discriminate between the two groups.

The principle can also be stated in terms of eigenvalues. Discriminant coefficients are chosen that maximize the eigenvalue for the composite variable, that is, the ratio of between-group to within-group sums of squares. A critical feature of these composite sums of squares is that they encapsulate, not only the variability of each variable, but also their *covariability.* This means that the coefficients are partial, just as in multiple and logistic regression, so each indicates the contribution of a particular variable while statistically controlling for all of the others. Further, the coefficients can again be calculated in unstandardized or standardized form, as in multiple regression. That said, we will see in the next section that discriminant coefficients are less informative than those in regression, whatever their form. After so many abstractions, it is more than time to return to an example where these ideas are made more concrete.

5.2.2 Two-Group Discriminant Analysis in Action

In order to appreciate similarities and differences in the techniques, it will be helpful to begin this section with a discriminant analysis that parallels the logistic regression carried out in Subsection 5.1.2. As a reminder, the data set consists of three independent variables: professional relations, autonomy, and control; and a dichotomous dependent variable: good mental health versus poor mental health. Bear in mind, though, that the independent/dependent status of these variables will flip occasionally in our discussion. This somersault is potentially confusing, but it does avoid deeper confusions that can arise in a nonstatistical account.

We begin with the question of whether the composite variable or discriminant function discriminates between the two groups to a degree that is unlikely to be due to chance. This is equivalent to asking whether the multivariate association between the independent and dependent variables is statistically significant. The null hypothesis that there is no multivariate association in the population can be tested using chi^2, which in the present case is 8.45 with a reassuring *p* value of .038. Since it is statistically significant it is meaningful to ask about the magnitude of the relationship. Discriminant analysis produces a multivariate version of Wilks's lambda (see Chapter 1), which has a value of .948 in this case. This means that the discriminant function or composite variable fails to account for 94.8% of the variance in mental health status. Conversely, by subtraction the function does account for 5.2% of variance. In the bivariate context this explained variance is indexed by the eta^2 statistic, but in this multivariate context it becomes known as the **canonical correlation^2**. So in the present analysis SPSS reports a canonical correlation of .228, that is the square root of .052. In summary, the discriminant analysis suggests that the three work variables considered as a set are related to mental health status and explain just over 5% of its variance. This outcome opens the way to an inspection of the coefficients in the composite variable to discover which variables are contributing to its discriminating power.

The contributions of individual variables can be shown in a variety of ways, and the most common appear in Table 5.3. The **unstandardized discriminant coefficients** in the first column are the weights used to generate the discriminant score. However, since they do not take account of any differences

Table 5.3 Three Types of Discriminant Coefficients Showing the Contributions of Three Workplace Variables to the Discriminant Function

Variable	*Unstandardized Coefficients*	*Standardized Coefficients*	*Structure Coefficients*
Professional relations	.407	1.126	.976
Autonomy	−.057	−.174	.375
Control	−.025	−.110	.305
Constant	−4.567		

in the measurement scales of the variable, they are not usually comparable, as in multiple and logistic regression. The **standardized discriminant coefficients** in the middle column are comparable, but only in a limited sense. Their rank order, ignoring the signs, provides an indication of the relative contribution made by variables to the discriminant function. Thus it is clear that the professional relations variable takes the lions' share of the credit in this case. Note that unlike beta coefficients in multiple regression, these coefficients cannot be interpreted in rate of change terms, nor do they have associated statistical tests. Since the standardized coefficients have been adjusted to take account of correlations among the variables, it is also helpful to have an unadjusted view of their contributions for comparison. This is provided by the **discriminant structure coefficients** in the last column. The structure coefficient is the simple correlation between scores on a particular variable and the discriminant scores. It thereby gives an uncluttered view of a variable's contribution and is favored by many analysts because of this. Ideally, as in the present case, the standardized and structure coefficients provide a similar message, although sometimes the differences can be instructive. The message is clearly that the professional relations variable is the key discriminator, as we found in the logistic regression analysis.

Another similarity to logistic regression is the availability of classification analysis: the prediction of group membership and the assessment of its success. In fact, in some discriminant analyses, particularly in applied settings, this is of more interest than the inspection of coefficients. There are a variety of ways to conduct classification analyses that can be pursued in the readings at the end of this chapter. In SPSS the discriminant scores are used to calculate what is called each case's **posterior probability:** their probability of being in a particular category given their discriminant score. This is then adjusted by the case's **prior probability:** the probability of their being in a category regardless of their discriminant score. The result of all this is a predicted group membership for each case, that is, the category that is their most probable location given their attributes. This predicted category can then be compared with their actual category to calculate various indices of classification success for the sample as a whole.

In the nurses' sample, about 84% of those in good mental health were correctly classified, and about 32% of those actually in poor health were so classified, giving an overall hit rate of nearly 61%. As we discussed in the context of logistic regression, these figures can and should be compared with what could be achieved by chance. However, as before, the notion of "chance" can take on a variety of meanings, and it is important to choose the most appropriate for the research context. In this form of classification analysis the choice of chance level is equivalent to the choice of prior probabilities. The simplest choice would be the tossed coin model, which gives a prior probability of 50% for either category. Using this would lead to the conclusion that the overall hit rate was 11% better than chance, but that this hid a gain of 34% in the good health group and a loss of 18% in the poor health group. Since the actual groups were not equal in size, a better choice of prior probabilities would be the sample relative frequencies: 56% in the good health group and 44% in the poor health group. Using these as base rates gives an overall hit rate gain of about 10% with a gain of 28% in the good health group and a loss of 12% in the poor health group.

This change in the choice of prior probabilities results in a relatively small shift in the pattern of classification success. However, there may be grounds for choosing quite different prior probabilities that could alter the rates considerably. For example, population figures may be available for the prevalence of mental health in nurses that are notably different from the sample figures and that might provide more accurate prior probabilities. Or there may be good reason to set the prior probabilities in a way that favors the accurate detection of mental health problems at the expense of detecting those in good mental health. Details aside, the two general points to reiterate from the earlier discussion on classification analysis in logistic regression are that the results hinge on the analyst's choice of prior probability, and that until the results are cross-validated in some way they should be regarded as how good classification can get.

The results of the discriminant analysis suggest that work characteristics can discriminate among nurses in good versus poor mental health, albeit to a modest degree. However, they also suggest that most if not all of this discriminative power is due to the professional relations variable. Since there are no individual variable tests in discriminant analysis, it is wise to check their contributions by running sets of analyses with them present or absent. If we run a discriminant analysis including only the professional relations variable, we find very similar figures to those resulting from the three variable model. The figures for the single variable model, with those from the three variable model in brackets, are $chi^2 = 8.12$ (8.45), $p = .004$ (.038), Wilks's lambda = .948 (.951), canonical correlation = .222 (.228), and overall classification success 62% (61%). The similarities here make it clear that the autonomy and control variables are quite redundant. Of course, all we have really done here is to repeat the ANOVA we conducted earlier. When there is only one independent variable, discriminant analysis collapses into an analysis of variance: a further demonstration of the cumulative nature of multivariate statistics. The example nonetheless exemplifies the value of comparing models with different subsets of variables to clarify their individual contributions.

To complete this section and to help consolidate understandings of the main features of discriminant analysis, we can turn to an example of a two-group analysis from the research literature on well-being. Philips and Murrell (1994) compared a group of 120 older adults who sought help for their mental health with another similar group of 120 who did not seek help, to see if they differed in terms of their well-being, experience of undesirable events, social integration, social support, and physical health. The discriminating power of these independent variables, accompanied by 10 sociodemographic control variables, was analyzed with a two-group discriminant analysis. The authors report the following statistics for the discriminant function: Wilks's lambda = .5861; $chi^2 = 122.89$, $p < .0001$; canonical correlation = .643. From the chi^2 and associated p value it is clear that the independent variable composite's capacity for discriminating between the two groups was highly unlikely to be due to chance. The extent of this capacity can be quantified by squaring the canonical correlation and concluding that the variables explained about 41% of the variance in help-seeking status. This same figure can be arrived at by subtracting the Wilks's lambda figure from 1 and converting to a percentage, since Wilks's lambda reflects unexplained variance.

To evaluate the contribution of individual variables, the authors present the standardized and structure coefficients for each. The rank orders of these two types of coefficient were strikingly different in some respects, reflecting the adjustments made to the standardized coefficients to account for correlations among the independent variables. Like many analysts, the authors chose to base their interpretations on the structure coefficients: the simple, unadjusted correlations between variable and discriminant scores. From these they concluded that the helpseekers had poorer well-being (structure coefficient = .77) and physical health (.63), experienced more undesirable events (.50), and reported less social support (.33). The social integration variable had a structure coefficient of .08, well below the conventional .30 threshold for interpretation (Hair, Anderson, Tatham & Black, 1998). Interestingly, they did not proceed to analyze a model without the sociodemographic variables. The full model suggested that these variables were contributing little, so it would have been interesting to evaluate the discriminating power of the five variables that were apparently making the sole contribution to predicting group membership. It is also interesting to note that the authors chose not to proceed to a classification analysis, despite the implications this might have had for mental health services for older adults.

5.2.3 Trustworthiness in Discriminant Analysis

In this section we review concerns about sampling, measurement, and the statistical assumptions that can influence the trustworthiness of discriminant analysis results. As in the case of logistic regression, the review can be relatively brief since it draws heavily on more extensive discussions in Section 4.3 of Chapter 4 and in Chapter 2.

It is generally recommended that the sample size in a discriminant analysis should provide at least 20 cases for each independent variable. A sample size smaller than this can result in discriminant coefficients that are not stable across samples and therefore not trustworthy (Stevens, 2002, p. 289; Hair et al., 1998, p. 258). It is also recommended that the smallest group size in the dependent variable categories be at least 20, with an absolute minimum greater than the number of independent variables. If these conditions are met, unequal sample sizes across the categories are not problematic in themselves, though they may have implications for the choice of prior probabilities in a classification analysis and may contribute to assumption violation as discussed below. As usual, the more general issues of acquiring enough cases to achieve appropriate levels of Type I and II error for the expected effect size, to compensate for unreliable measurement, to allow for missing data, and to create a holdout sample if desired, should all be considered in determining the optimum sample size.

The dependent variable in a discriminant analysis should be categorical and may have any number of categories. The categories should be mutually exclusive and jointly comprehensive, allowing each case to be assigned to a single category. It is assumed that all of the independent variables are measured on at least an interval scale. Including other types of variables using dummy coding will produce meaningful results. However, the more noninterval variables that are included, the less trustworthy the results will be in terms of finding the optimum separation of the groups. In this situation it is usually wiser to resort to logistic regression, which can accommodate any mix of independent variable types. No new issues of measurement quality arise in discriminant analysis; as usual, reliable and valid measurement of all variables is the order of the day.

The statistical assumptions required for discriminant analysis are essentially the same as for OLS regression, though some of them take on a slightly different form. The independent variables are assumed to have a multivariate normal distribution in each population from which the category samples are drawn. As in OLS regression, the consequences of violating this assumption are not usually serious if the sample size requirements above are met. The assumption of multivariate homoscedasticity found in OLS regression takes on a more elaborate form in the present context. Discriminant analysis requires that the population variances *and covariances* for all independent variables are equal across the dependent variable groups. This is known as the **homogeneity of variance-covariance matrices** assumption. The status of the assumption can be explored by inspecting the group variances and covariances, examining appropriate plots, and testing with statistics such as **Box's M.** If the sample sizes in each category are reasonably large and approximately equal, violation of this assumption has little effect on statistical tests, but classification analyses may be distorted. If there is a clear violation, remedies may be found in transformations of variables, or possibly in an alternative approach to classification called quadratic discrimination.

Discriminant analysis also assumes independence of cases and multivariate linearity of relationships among the independent variables in each category of the dependent variable. As in OLS regression, multicollinearity or high correlations among independent variables can be a problem to which the analyst should be alert. Also, outliers on the independent variables may distort the results. Examination of univariate and bivariate statistics and plots for the independent variables is important to check for these potential problems and for nonnormality. As we noted earlier, it is also possible to explore the multivariate structure of the independent variables by treating one of them as a pseudodependent variable and conducting an OLS multiple regression. This will generate all of the diagnostic tools that were discussed in Chapter 4 that can now be used to evaluate many of the assumptions required by discriminant analysis.

5.2.4 Extending the Scope of Discriminant Analysis

In Subsection 5.1.4 we discussed how the scope of binary logistic regression could be extended by accommodating various types of independent variables, conducting sequential analyses, and analyzing dependent variables with more than two categories or groups. As we noted earlier, it is not advisable in discriminant analysis to include independent variables that have less than interval scaling, so there is no need to pursue that topic here. All of the sequential strategies, both hierarchical and statistical, can be used in discriminant analysis, though the statistical approach using such techniques as stepwise analysis is the most common application. No new general issues arise when sequential strategies are used in discriminant analysis, so the earlier discussion in Chapter 4 on OLS regression should suffice as an introduction. This then leaves the topic of how discriminant analysis can be applied to a dependent variable with more than two groups.

When discriminant analysis is applied to more than two groups, the major consequence is that more than one discriminant function can be calculated. Each function will have its own set of coefficients and each will generate a discriminant score for every case. Mathematically, it is possible to derive as many functions as there are groups minus 1. So for a four-group analysis, there will be a maximum of three functions, and each case will potentially have three discriminant scores. However, the fact that three functions can be derived does not mean that all are necessary in order to achieve maximum discrimination between the groups. This may be achievable with only one, or perhaps two, of the available functions. Not surprisingly then, the major new issue that arises when the dependent variable has more than two categories is how many functions are worth retaining from those that are available.

The broad strategy for deriving and testing multiple discriminant functions can be confusing, so we will first describe the logic in broad outline and then make it less abstract with a closing research example. When there two groups, there is only one dimension along which their two means on the composite variable (centroids) can be pushed apart as far as possible. However, each time a group is added, another dimension emerges along which the centroids can be separated. A four-group discriminant analysis, for example, first calculates the set of coefficients that maximally separates the four centroids: the first discriminant function. It then calculates another set of coefficients that separates the centroids in a completely different way. This is the second discriminant function, whose discriminatory power is unrelated to that found in the first function. The process then continues to derive the third function. Since the objective is to maximize the discriminating power of a function, the result is a series of functions (three in this case) that have decreasing discriminating power and that are uncorrelated with each other.

Another way of thinking about multiple discriminant functions is in terms of explained variance. All of the available functions in an analysis are jointly responsible for any explained variance that is achieved. Deriving separate functions can be seen as assigning portions of this explained variance to each function. The first function will be awarded the largest portion, and the succeeding functions will receive diminishing portions. Moreover, the portions will be mutually exclusive, so that they add up to the total explained variance. In general then, a discriminant analysis will produce a series of functions one less in number than the number of groups or categories in the dependent variable. These functions will be ordered such that they have decreasing discriminating or explanatory power, and each will achieve this power in different ways.

As we noted, the question this creates is whether all of the functions are worth retaining in the analysis. The usual approach to this question is to rely on the statistical significance of functions. Unfortunately, this is not simply a matter of testing the significance of each function. The first step is to test the significance of all the available functions considered jointly. This makes sense, as it is equivalent to testing whether the functions jointly capture more explained variance than would be expected by chance. If this test achieves statistical significance, the way is open to testing for superfluous functions. (If not, the analysis is not worth pursuing at all.) The superfluity tests proceed by testing the significance of subsets of functions, each time omitting the next largest function. So, in the three-function case, the first test would evaluate the joint significance of the second and third functions; and the second test would evaluate the significance of the third function alone. The occurrence of a statistically *non*significant result suggests that none of the functions under test at that point are worth retaining. To see this process in action, we can turn to a study involving four groups, which meant that three functions were available in principle.

Diehl, Elnick, Bourbeau, and Labouvie-Vief (1998) conducted a study to examine how adult attachment styles are related to a range of family context and personality variables. They identified 304 cases as exhibiting one of four attachment styles: secure (154), dismissing (77), preoccupied (25), or fearful (48). One of their research objectives was to discover how well a wide range of well-being, family, and personality variables would predict membership of these four groups. To find out, they conducted a four-group discriminant analysis with 12 independent variables. Since there were four groups, it was possible to derive three functions. Their first analytic task was to decide whether three functions were needed to account for any discriminating power of the independent variables, or whether a smaller number would suffice.

All three functions had a joint chi^2 of 109.96, with an associated p value $< .001$. This demonstrated an overall level of discriminatory power that was unlikely to be due to chance. The reported Wilks's lambda of .69 indicates that the functions accounted for $100(1 - .69) = 31\%$ of the variance in attachment styles. The next test of functions 2 and 3 was also significant with a chi^2 value of 47.24 and a $p < .01$. This outcome, and the fact that the second function accounted for about 13% of the variance in attachment styles, suggested that it was worth retaining for its discriminating power. However, when the third function was tested it did not achieve statistical significance and accounted for a minuscule amount of variance in the dependent variable. Moreover, functions 1 and 2 together accounted for over 95% of the explained variance (not of the total variance). All of this suggested that only two functions were required rather than the three that were available in principle.

Since two functions were retained, two sets of coefficients resulted, and the authors present the two sets of structure coefficients as the basis for their interpretations of how the independent variables contribute to the functions. Based on the patterning of the coefficients, the researchers labeled the first function as a "self-model" and the second function as an "other-model." They also examined the group means (centroids) on the two discriminant scores and found that the first function discriminated the secure and dismissing styles from the preoccupied and fearful styles. In contrast, the second function discriminated the secure and preoccupied styles from the dismissing and fearful styles. This differential outcome demonstrates nicely how the two functions captured different aspects of the discriminating power of the super function.

Finally, Diehl et al. (1998) conducted a classification analysis. Given the very different numbers of cases in each category, they wisely chose the relative frequencies as their prior probabilities rather than a uniform 25% for each category. The success rates for each category with the prior probability in brackets were secure 50% (51%), dismissing 55.8% (25%), preoccupied 44% (8%), and fearful 50% (16%). The overall success rate for the classification analysis was 51%. As the authors note, the prediction gains from using the discriminating power in the independent variables were over 30%, but only in the groups with insecure styles.

5.3 FURTHER READING

Pampel's (2000) "primer" on logistic regression is exactly that—a model of clear exposition for the novice, while Tabachnick and Fidell (2001, Chapter 12) provide a more extensive, computer-analysis-oriented account. In the present context, the chapter by Hair et al. (1998, Chapter 5) is particularly germane because it discusses logistic regression and discriminant analysis in parallel. After more than 20 years, Klecka's (1980) brief introduction to discriminant analysis remains a valuable source. More extensive treatments of discriminant analysis can be found in Tabachnick and Fidell (2001, Chapter 11), Stevens (2002, Chapter 7), and Huberty (1984, 1994).

SIX

MULTIVARIATE ANALYSIS OF VARIANCE

Whatever their differences, the techniques of multiple regression, logistic regression, and discriminant analysis that were discussed in Chapters 4 and 5 share a capacity for analyzing the effects of multiple independent variables on a *single* dependent variable. Multivariate analysis of variance (MANOVA) is distinctive in that it allows multiplicity of both independent and dependent variables. In fact, it is the first true multivariate technique that we have encountered, as strictly speaking, the "multi" in "multivariate" refers to multiple dependent variables. To make sense of MANOVA, it is helpful to approach it through its simpler relative—analysis of variance (ANOVA)—which allows only one dependent variable. In Chapter 1 we discussed the most basic form of ANOVA, which has one independent variable with two categories. Our general strategy in this chapter will be to explore first how ANOVA can accommodate more complex independent variable structures. Having established these fundamentals regarding independent variables, we can then relocate them inside the MANOVA framework and show how they operate when there are multiple dependent variables. This shift into MANOVA will be relatively straightforward, not least because we have already encountered it in the guise of discriminant analysis in Chapter 5.

In Section 6.1 we review the basics of ANOVA, using an independent variable with three categories and a single dependent variable measured on an interval scale. If such a categorical variable is found to account for differences in the dependent variable, the question arises of which particular categories are implicated. This question can be explored using contrast analysis in which differences between particular categories are analyzed. In Section 6.2 we extend the discussion on ANOVA to so-called factorial designs, in which there are two or more independent variables or factors. The key idea here is that once there is more than one independent variable, an effect may be found for a single factor—a main effect, or for factors operating jointly—an interaction effect. We have encountered this idea before under the heading of moderated regression analysis. In fact, such analyses can be conducted with either ANOVA or multiple regression. Section 6.3 turns to MANOVA proper and shows how main and interaction effects can be analyzed when there are multiple dependent variables. As usual, the strategy will be to form the multiple variables into a composite, which is subsequently unpacked if an overall effect is found.

In Section 6.4 we turn to a different form of ANOVA/MANOVA. All of the analyses in Sections 6.1–6.3 require that the subgroups of cases in the categories of the independent variable(s) are independent of each other. These are known as **between-subjects analyses.** If this requirement is not met, the calculations will be erroneous and should be replaced by **within-subjects analyses.** Probably the most common reason for using within-subjects analyses is the presence of one or more **repeated-measures factors.** As the name suggests, this means that the same cases appear in more than one category of an independent variable, for example, if they provide the same measurements on more than one occasion. Traditionally, this type of data has been analyzed with a special form of ANOVA, but this is legitimate only under certain restrictive assumptions. The solution is to recast the ANOVA as a MANOVA that has less restrictive assumptions. Section 6.4 provides a discussion of within-subjects analysis using ANOVA and MANOVA.

Section 6.5 explores the usual issues of trustworthiness in the context of ANOVA/MANOVA. This will be a short section as virtually all of the issues have been reviewed in earlier chapters. Finally, Section 6.6 discusses an elaboration of analysis of variance, called **analysis of covariance.** This extension allows for the inclusion of interval-level independent variables alongside the categorical factors so that the effects of the former can be statistically controlled. When there is one dependent variable, this technique is called analysis of covariance (ANCOVA); when there are multiple dependent variables, the technique is called multivariate analysis of covariance (MANCOVA).

6.1 ONE-WAY ANALYSIS OF VARIANCE

One-way analysis of variance is used to analyze the relationship between one categorical independent variable and one interval-level dependent variable. The term "one-way" refers to the singular independent variable. Although, true to its name, the analysis revolves around variances, it is easier to think of ANOVA, and subsequently MANOVA, as techniques for analyzing sets of differences between mean scores on the dependent variable(s). However complex the analysis becomes, differences between means, and therefore *group* differences, always remain the focal point. This will become apparent as we return to the nurses' data and analyze them further, first with ANOVA and then with MANOVA.

First we return to the question of whether the quality of the nurses' professional relations has an effect on their mental health. For present purposes, we will treat professional relations as a categorical independent variable with three categories: poor, moderate, and good. The dependent variable is their scores on the mental health measure. With this data structure, any relationship between the two variables will be evident in differences in mean mental health score across the three categories. The three means are poor 66.7, moderate 73.6, and good 77.9, and the grand mean is 72.6. This pattern of mean differences suggests that better mental health scores are associated with better professional relations. But how can we be confident that this pattern is not due to chance?

In Chapters 1 and 2 we discussed how ANOVA, revolving around the F statistic, can be used to test the null hypothesis that two population means are identical; in that instance, that women and men reported the same level of happiness. As we noted then, the statistical significance of a difference between two means can also be tested with a t statistic, but the ANOVA approach has the advantage that it can be applied to the differences between any number of means. In the present example the F statistic can be used to test the null hypothesis that the three means are identical in the population sampled. If this "omnibus" null hypothesis is rejected, it suggests that at least one of the mean differences is unlikely to be due to chance.

To review the fundamentals of ANOVA, it will be helpful to examine the contents of the summary table reporting the results of this particular analysis, shown in Table 6.1. The between-group sum of squares in the top row captures the mean differences in mental health across the three professional relations categories. At the heart of this calculation is the difference between each mean

Table 6.1 One-Way ANOVA Showing the Relationship Between Professional Relations and Mental Health

Source	*Sum of Squares*	*df*	*Mean Square*	*F*	*p*
Between groups	3,254.51	2	1,627.26	6.98	.001
Within groups	37,319.13	160	233.25		
Total	40,573.64	162			

and the grand mean, that is, the between-groups variability. The greater this variability, the more evidence there is that professional relations and mental health are associated. This sum of squares is then converted into a variance or mean square by dividing it by the appropriate degrees of freedom (*df*). The degrees of freedom here are equal to the number of data points (i.e., groups) minus 1, that is, $3 - 1 = 2$, and so the mean square is $3{,}254.51/2 = 1{,}627.26$.

This between-groups variability is compared with the within-groups variability, that is, the extent to which these individuals differ in mental health, regardless of their professional relations category. At the heart of this calculation is the difference between each individual score and the mean for their group or category. All of this within-groups variability is summed or pooled across the three groups to derive the within-groups sum of squares. This, too, is divided by its degrees of freedom to calculate the within-groups mean square or variance. The degrees of freedom for each group is one less than the number of data points. Since there are 163 cases in all, this means that the within-groups degrees of freedom is $163 - 3 = 160$, and the within-groups mean square is $37{,}319.13/160 = 233.25$.

We now turn to the analysis of the variances as such. The ratio of the between-groups to the within-groups variances (mean squares) gives the *F* ratio of 6.98. With the degrees of freedom shown in the Table 6.1, this *F* value has an associated *p* value of .001. This means that if we have adopted a conventional alpha level of .05, we can safely reject the null hypothesis of no differences among the population means: They are clearly not equal in some way, and this inequality is highly unlikely to be due to chance. At this point then it seems that there is evidence of an association between professional relations and mental health in this sample of nurses.

When we used ANOVA to test the difference between two means in Chapter 2, the story ended at this point. But once we have more than two means, a further issue inevitably arises. Where exactly are the inequalities in the set of mean differences, and how can we test any such differences while still maintaining our chosen alpha level? Remember that p value calculations are based on the assumption of a single analysis. As the number of analyses involving the same means accumulates, the effect is to inflate the Type I error rate. Detecting specific differences and controlling alpha inflation takes us into the realm of **contrast analysis.** This is a complex area that generates a variety of problems and opinions. This fact, and limitations of space, means that we can do no more than hint at its potential.

The first point to appreciate is the sheer number of comparisons that are possible, even when only three means are available. To recap, in the present example the mental health means are 66.7 for nurses with poor relations, 73.6 for the moderate relations group, and 77.9 for those with good relations. Clearly, we could test the three pairwise differences, but also we could test the differences between subsets: poor against the average of moderate and good, and so forth. We also could test for trend, most obviously for a positive linear trend across the three means. This potential for many comparisons leads to a fundamental distinction that guides their analysis. Comparisons are divided into those that are decided on when the study objectives are defined, and those that are chosen after the omnibus test has indicated that the population means are unequal. The former are called **planned comparisons,** and the latter are referred to as **post hoc comparisons.**

If the analyst has planned specific comparisons across the categories of an independent variable, the outcome of the omnibus test is typically of no interest or consequence in itself. In this situation, the study has been planned to test specific hypotheses about category means rather than the more diffuse hypothesis evaluated by the omnibus test. In fact, as Rosenthal and Rosnow (1985) demonstrate, a statistically insignificant omnibus test can be compatible with a statistically significant comparison in some situations and therefore misleading if used as a starting point. When there are just a few multiple planned comparisons to be analyzed, the inflating alpha problem is generally thought to be of no consequence, especially if the omnibus test is statistically significant. However, if there are more than a few, a **Bonferroni adjustment** to deflate the alpha is usually recommended. (Keppel [1991] recommends such an adjustment if there are more comparisons than between-groups degrees of freedom, that is, more than two comparisons in the present example.) In its simplest form, the adjustment involves dividing the chosen alpha by the number of comparisons and adopting the result as the alpha for comparison tests. For example, if the chosen alpha were .05 and we planned to conduct five comparisons, the p value for each comparison would be compared with an alpha of $.05/5 = .01$.

When comparisons are post hoc, as in the present example, there is a wide range of strategies for dealing with the inflating alpha problem. These strategies vary according to the type of comparison envisaged and to the severity of the deflation they produce. At the gentle end of the spectrum is the zero deflation strategy, which is seen as allowable if the omnibus test is statistically significant. In the middle of the severity spectrum are tests such as the memorably named **Tukey's honestly significant difference test.** The prize for maximum severity goes to **Scheffé's test.** We can see the differential effects of these tests by applying them to the pairwise differences among our three means. The unadjusted comparisons, justified by the outcome of the omnibus test, show that the poor versus moderate ($p = .02$) and poor versus good differences ($p < .001$) are statistically significant, while the moderate versus good difference ($p = .15$) is not. The corresponding p values when the Tukey test is applied are .04, .001, and .31, respectively. So, although the p values have been increased (i.e., the alpha has been reduced), the pattern of outcomes is the same as for the unadjusted tests. However, when the Scheffé test is applied, only the poor versus good difference remains significant, reflecting the more severe deflationary adjustment.

What should we conclude from these post hoc analyses? The Scheffé test is not recommended when the omnibus test is statistically significant, or when simple pairwise comparisons are undertaken. So it seems safe to conclude that nurses with poor professional relations have worse mental health than either those with moderate or those with good relations. We are not justified in claiming a difference in mental health between those with moderate professional relations and those with good professional relations.

As we noted earlier, this account just scratches the surface of a complex and contentious area. To illustrate the contentiousness, we can finish this discussion of contrast analysis with a very different view. Kenneth Rothman, an eminent epidemiologist, charmingly reduces the issue of comparison control strategies to absurdity as follows:

> If many comparisons were made and each one were reported individually, let us say in a separate publication, it would be absurd to make adjustments to the reported P-values in each report based on the total number of reports. If such adjustments were indicated, it would also follow that an investigator

> should keep a cumulative total of comparisons made during a career, and adjust all "significance" tests according to the current total of comparisons made to date. The more senior the investigator, the more the *P*-value would have to be inflated. (Rothman, 1986, pp. 148–149)

The point of this brief discussion is simply to raise the issue of multiple tests, not to resolve it. Analysts have to decide how best to maximize the trustworthiness of their results in this respect. At the very least, most would agree with Rothman that the most fundamental requirement is to report clearly how many tests have been conducted, whether or not any adjustments have been made.

An example of the use of one-way ANOVA with post hoc comparisons can be found in a study by Mookherjee and Harsha (1997) of the relationship between marital status and perceived well-being in a national sample of 12,168 adults in the United States. This study is also an interesting example of how a large sample size can result in small mean differences that are statistically significant. To help counter this problem, Mookherjee and Harsha cautiously adopted an alpha of .001 rather than the more conventional .05. They reported mean well-being scores for five different categories of marital status as follows: married 14.91, widowed 14.32, divorced 13.73, separated 13.18, and never married 14.06. (The score range for the whole sample was 6–18, with a mean of 14.48 and standard deviation of 2.50.) A one-way ANOVA produced an F of 136.6 with an associated p value less than .0001. Following this statistically significant omnibus test, Mookherjee and Harsha conducted five pairwise comparisons and a comparison between the married group and all other groups combined. All of these unadjusted tests were statistically significant with p values less than their chosen alpha of .001. Although these tests are defensible, the interpretation of some of them suggests some confusion between statistical and substantive significance. For example, the contention that a significant difference between the married and widowed groups indicates that partner loss has "a substantial negative effect on perception of well-being" (p. 98) seems hard to justify on the basis of a mean difference of 0.6 on a 12-point scale.

6.2 FACTORIAL ANALYSIS OF VARIANCE

Continuing with the nursing example, we have concluded that professional relations are associated with mental health, but maybe the picture is more complicated than this. Maybe the association is confounded by other factors, or maybe the association varies across different subgroups of nurses—an interaction or moderating effect. For example, it is possible that the number of years' experience as a nurse may be a confounding variable, producing both better professional relations and better mental health. Additionally, or alternatively, professional relations and mental health may be associated differently according to length of experience. Perhaps poor relations matter less to more experienced nurses and so are less likely to affect their mental health.

As soon as another independent variable is introduced, we need **factorial ANOVA** to answer questions about individual (main) and joint (interaction) effects and about possible confounding. The term "factorial" indicates the presence of more than one independent variable, with the suffix "way" signaling their number. So, for the present example, we are about to conduct a two-way factorial ANOVA with mental health as the dependent variable. The professional relations factor will continue to have three categories (poor, moderate, and good), and the experience factor will have two categories ($\leq$ 10 years' and > 10 years' experience). So we can describe the analysis more precisely by saying that it has a 3×2 design, thereby encapsulating both the number of factors and the number of categories in each.

As we noted at the outset, however complicated an analysis of variance becomes, it always addresses a set of means. The means for the present analysis are shown in Table 6.2.

The mental health mean for all 163 cases—the grand mean—appears in the bottom right corner of the table. The means for the categories in each of the two independent variables are shown along the outside or margins of the table. So the row marginal means along the bottom show the effect for length of experience, and the column marginal means on the far right show the effect for professional relations. The means in the body of the table—the cell means—are those for each combination of categories. Since there are 3×2 cross-categories, there are 6 cell means that show what happens when the two independent variables are combined. In short, the marginal means represent the main effects and the cell means represent the interaction effect (though we will need to modify this statement a little in a moment).

What does the patterning of the means suggest about the effects of professional relations and experience on mental health? The column marginal means are familiar and show an apparent enhancement of mental health for nurses with better relations. Along the bottom of the table, the row marginal

Table 6.2 Mean Mental Health Scores by Categories of Professional Relations and Length of Experience

	Length of Experience		
Professional Relations	*≤ 10 Years*	*> 10 Years*	*Total*
Poor	62.5	71.1	66.7
Moderate	73.9	73.4	73.6
Good	74.5	80.1	77.9
Total	70.0	74.9	72.6

means suggest that nurses with more experience enjoy better mental health: almost a 5-point difference on average. The cell means read columnwise also suggest that experience may be moderating the effect of professional relations, that is, an interaction effect. Thus, it appears that for nurses with less experience, mental health differs notably between the poor and moderate groups. But for more experienced nurses, the difference is more striking between the moderate and good groups. In other words, the nature of the association between professional relations and mental health differs according to the level of experience.

This type of exploration is useful, but it can be error prone unless accompanied by a statistical analysis such as ANOVA, for a number of reasons. The first is that while the table shows group differences, it gives no indication of *individual* differences. The groups are made up of different individuals, and it is possible that this differential composition is sufficient to account for the group differences. This is precisely the issue that ANOVA is designed to address by manipulating between-groups and within-groups variability. The second reason is that the effects of professional relations and experience may be confounded with each other, and this cannot be gauged from the observed means. In the present example the cell means are based on different-sized subsamples ranging from 19 to 32. This is called an **unbalanced design,** which is fertile ground for confounding since it effectively creates a correlation between the independent variables. To counter this, the ANOVA sums of squares for an unbalanced design need to be calculated in a regression style, adjusting for correlations among the variables.

Finally, it is possible that the interaction effect suggested by the cell means is more apparent than real. This is partly because of the first point about the need to allow for individual differences. But in addition, as Rosenthal and

Table 6.3 Two-Way ANOVA Showing the Effects of Professional Relations and Experience on Mental Health

Source	*Sum of Squares*	*df*	*Mean Square*	*F*	*p*
Professional relations	2,871.94	2	1,435.97	6.27	.002
Years' experience	823.51	1	823.51	3.59	.060
Interaction	615.26	2	307.63	1.34	.264
Error	35,983.74	157	229.20		
Total	40,573.64	162			

Rosnow (1985) have pointed out, the cell means represent the main and interaction effects combined. Since an interaction is thought of as a joint effect of variables *over and above* their individual effects, the latter need to be removed if a clear picture of the pattern of interaction is to be gained. Rosenthal and Rosnow show how this can be achieved, but the details lie beyond the scope of the present discussion. For now, it is sufficient to note that an interaction effect should always be tested and, if shown to be statistically significant, should then be interpreted with care.

All of this makes it timely to turn to the results of the two-way ANOVA of the nurses' data. These are shown in the conventional format in Table 6.3.

The bottom row showing the variability in mental health for the whole sample is the same as in the earlier one-way analysis. This reflects the fact that however complex the analysis with respect to the independent variables, the general objective remains that of accounting for differences in the dependent variable. In the one-way analysis the effect of the single independent variable was represented by the numbers in the row labeled between-groups. We are still interested in between-groups variability, but now we want to divide it into three portions representing the two main effects and the interaction effect, respectively. Note that, since there are $3 \times 2 = 6$ groups, the degrees of freedom for the three effects add up to $6 - 1 = 5$. The degrees of freedom for each main effect are the number of categories minus 1, while the interaction degrees of freedom are the product of those for the main effects. The sums of squares for each of the three effects can be calculated in various ways, depending on what sort of statistical control is thought desirable. In the present analysis they have been calculated so that each main effect is controlled for the other, and the interaction effect controls for the main effects. Deciding on the appropriate sum of squares calculations can be a complex affair, and a helpful discussion can be found in Stevens (2002, Chapter 8).

Information about individual differences within groups is contained in the row labeled "error." This pools the variability around the mean of each of the six groups to show how much individuals differ in mental health regardless of their status on the independent variables. Since there are six groups, each with n–1 degrees of freedom, the pooled degrees of freedom for error are 163 – 6 = 157. Each of the sums of squares is divided by its degrees of freedom to obtain the mean squares or variances shown in the third column of numbers. Then an F ratio is calculated for each of the three effects by dividing the effect mean square by the error mean square. Finally, a p value is calculated for each F, which can be compared with the chosen alpha.

If we assume a conventional alpha of .05, the p values in the table suggest that only professional relations have an effect on mental health. There is no main effect for experience, nor does it interact with professional relations. (It is interesting to note in passing that a one-way ANOVA of experience produces a marginally significant effect on mental health [$p = .046$]. However, this now becomes nonsignificant when it is adjusted for its correlation with professional relations, and when the error term for six rather than two groups is used.) The absence of an interaction effect means that the main effects can be interpreted independently of each other. It is important to note that had there been a statistically significant interaction effect, this would signal that each effect should be interpreted with reference to the other. Given two independent variables, the main effect of one can be thought of as having been averaged across the categories of the other. This averaging process is not misleading as long as the nature of the main effect does not change across the categories of the second variable. But the presence of an interaction tells us that there is such a change and that we need to examine the main effects jointly to uncover their pattern of interaction.

To exemplify the use of factorial ANOVA and interaction effects, we can return to the study by Mookherjee and Harsha (1997) on the determinants of perceived well-being. Earlier we focused on the relationship between marital status and well-being, but this study also included the factors of gender, race, and financial status. The most complex of the analyses consisted of a four-way ANOVA with four main effects: marital status (2 categories), gender (2 categories), race (2 categories), and financial status (3 categories); and all possible two-way, three-way, and four-way interactions among these factors. Thus the analysis had a $2 \times 2 \times 2 \times 3$ factorial design. Notice how the number of possible interaction effects explodes as the number of factors increases. In this case, four factors produce 6 two-way, 4 three-way, and 1 four-way interaction effects. This type of complex design can require large numbers of cases and can generate interpretive problems when higher-order interactions are detected, especially if they are not predicted. In the present example the very large sample size and near absence of interaction effects led to a simple interpretable outcome.

All four main effects, and the interaction between marital status and gender, were statistically significant with a p value less than .0001, well below the chosen alpha of .001. White respondents reported greater well-being than nonwhites, as did respondents with better financial status. The presence of the interaction between marital status and gender meant that their main effects could not be considered independently of each other. The nature of this interaction is not entirely clear from the information provided. However, it appears to show that while married respondents overall enjoyed greater well-being than nonmarrieds, this difference was more pronounced in men than in women. So the main effect for marital status is not misleading in this example; rather it averages across some interesting variation by gender. A similar pattern appeared when the interaction was approached from the perspective of gender. Women appeared to report higher levels of well-being than men, but this difference was greater in the nonmarried group.

It should be clear by now that ANOVA is a powerful form of analysis that can deal with multiple independent variables, each with as many categories as desired. Moreover, it can be used to detect not only the individual effects of each independent variable, but also their joint effects, while exercising statistical control of any correlations among the factors. When effects are found, they can be analyzed in more detail using contrast analysis. This can be used to look inside main effects involving factors with more than two categories, or inside complex interactions. Not surprisingly, given our discussion of dummy variables in Chapter 4, all of this can also be accomplished using multiple regression. In fact, it has long been recognized that ANOVA is a special case of regression (Cohen et al., 2003). The tools available inside each technique allow for different insights into the data under analysis, but the overall outcomes of hypothesis tests will be identical.

6.3 MULTIVARIATE ANALYSIS OF VARIANCE

To see how ANOVA can be elaborated into MANOVA, we will repeat the analysis of the effects of professional relations and experience in the nurses' data, but now include a second *dependent* variable. So the analysis will now focus on the main and interaction effects of two factors: professional relations (3 categories) and years' experience (2 categories), on the dependent variables of mental health and general health perception. As the name implies, the general health perception measure assesses respondents' overall evaluation of their health and illness status. Scores on this measure are scaled the same as those on the mental health measure so that they may fall within a 100-point range.

It is the introduction of a second dependent variable that leads to the use of MANOVA rather than two separate ANOVAs, one for each of the health variables. The advantages of MANOVA over multiple ANOVAs are the same as those we reviewed for all multivariate techniques in Chapter 3. MANOVA helps to avoid the inflating alpha problem by testing a set of dependent variables simultaneously rather than one at a time. This is not a major issue in the present circumstance where we have only two dependent variables, but it can quickly become so as their number increases. It also allows for the usual multivariate adjustments to control for correlations, now among the *dependent* variables. In the present data set, mental health and general health scores are correlated .43 ($p < .001$), so separate analyses of their determinants may well be confounded. Finally, there may be good reason to examine the effects of factors on a coherent set of variables that jointly map onto the content of a complex construct. There are clearly sound reasons for preferring MANOVA to a series of ANOVAs under some circumstances. However, this preference needs to be balanced with the understanding that an explicit rationale is needed to justify treating multiple dependent variables as a set and that a MANOVA may provide less statistical power than a series of ANOVAs. What is gained in multivariate statistical control may be lost in the failure to detect specific effects on particular dependent variables.

As noted earlier, the focus of any ANOVA or MANOVA is the pattern of mean differences on the dependent variable(s) across categories of the independent variables. Table 6.4 shows the means for the present MANOVA analysis. This analysis is based on 161 rather than 163 cases since two nurses failed to complete the general health measure. Accordingly, the mental health means are slightly different from those in Table 6.2.

Examination of the means for general health in the final column suggests that the quality of professional relations affects mental health and general health in a similar way. In contrast, the means in the bottom row suggest that experience makes less of an impact on general health than on mental health. The cell means for general health again seem to hint at an interaction effect. For the less experienced nurses, general health improves across each category

Table 6.4 Mean Scores on Mental Health and General Health by Categories of Professional Relations and Length of Experience

	Length of Experience					
	≤ 10 Years		*> 10 Years*		*Total*	
Professional Relations	*Mental Health*	*General Health*	*Mental Health*	*General Health*	*Mental Health*	*General Health*
Poor	62.5	66.5	70.9	74.2	66.5	70.2
Moderate	73.9	76.3	73.4	73.3	73.6	74.8
Good	74.0	82.2	80.1	81.4	77.8	81.7
Total	69.8	74.2	74.9	76.3	72.6	75.3

of the professional relations variable. But for the more experienced nurses, the general health of those with moderate relations is actually slightly worse than that of those with poor relations, with an improvement only evident for nurses with good relations. Conclusions reached from this type of inspection are fraught with danger for the reasons we reviewed earlier, but also because we are now comparing means across correlated dependent variables. To gain a more reliable picture, we need to turn to MANOVA.

In factorial ANOVA we use an F statistic to test a null hypothesis for each effect. For the example in Section 6.2, the F for the main effect of professional relations tested the null hypothesis that the three category means were equal in the population sampled; the F for the main effect of experience tested the null hypothesis that the two category means were equal; and the F for interaction tested the null hypothesis that there were no differences among the six cell means beyond those due to the main effects. A MANOVA analysis follows the same pattern of testing effects, but now each effect applies to a *set* of dependent variable means: a set of two in our example. So, for example, the multivariate test of the professional relations effect will test the null hypothesis that the *pairs* of dependent variable means are equal across the three categories of that factor in the population.

6.3.1 The Composite Variable in MANOVA

To enable this setwise testing, we make use of the usual multivariate strategy of forming a composite variable for each effect, but on this occasion a composite of the *dependent* variables. The form of this composite will look familiar if you have read the material on discriminant analysis in Chapter 5. As we noted at the time, discriminant analysis is identical to MANOVA except for the reversed status of the independent and dependent variables. The general form of the composite variable for MANOVA is:

discriminant score = [discriminant coefficient 1(DV1)
+ discriminant coefficient 2(DV2) + ...
+ constant]

There is a different composite variable for each main and interaction effect in the MANOVA design. To see how the composite works, we will focus on the main effect for professional relations. A discriminant score can be computed for each case by adding their weighted scores on the two dependent variables, as shown in the equation. These discriminant scores can be divided up into the poor, moderate, and good relations groups, and the mean discriminant score can be calculated for each of the three groups. As we noted in Chapter 5, the group means on the composite variable are known as **centroids.** The discriminant coefficients are chosen so that the distance between the centroids is maximized. In other words, they are calculated in such a way as to push the three group means on the composite variable as far apart as possible. Each main and interaction effect has its own composite variable, so if we were calculating the discriminant scores for the experience main effect, there would be different coefficients, derived by maximizing the distance between the group centroids for the two experience groups.

As we noted in Chapter 5, the criterion for forming composites can also be stated in terms of eigenvalues. Discriminant coefficients are chosen that maximize the eigenvalue for the composite variable, that is, the ratio of between-group to within-group sums of squares. A critical feature of these composite sums of squares is that they encapsulate not only the variability of each dependent variable, but also their covariability. This means that the coefficients are partial, so each indicates the effect on a particular dependent variable while statistically controlling for correlations among all of the dependent variables.

6.3.2 MANOVA in Action

The discriminant scores can be entered into ANOVA-style calculations to produce a *multivariate F* statistic for each effect. These MANOVA calculations

Table 6.5 MANOVA Results Showing the Effects of Professional Relations and Length of Experience on Mental Health and General Health

Effect	*Wilks's Lambda*	*F*	*p*
Professional relations	.909	3.74	.005
Years' experience	.976	1.88	.156
Interaction	.975	.98	.421

are more complex than in ANOVA because, as we just noted, they include manipulations of covariability as well as of variability, but the underlying logic is the same. A MANOVA of the data in our example produced the results shown in Table 6.5.

The first point to reemphasize about these results is that the test of each of the three effects is based on a different composite variable. As we noted, the coefficients in each composite are derived by maximizing the distances between *different* subsets of centroids. The second point to reemphasize is that these simple numbers hide the considerable amount of statistical control that is being exercised in the background on the correlations among the independent variables and among the dependent variables. Thus the main effects are controlled for each other, and the interaction effect is controlled for the main effects. Further, as we noted, the calculations also control for the correlation between mental health and general health. All of this is achieved by adjusting between-groups sums of squares for each effect. Fortunately, we can avoid the detail of how this is accomplished and just admire the outcome.

The eigenvalues can be manipulated in various ways to derive multivariate statistics that can in turn be tested with the *F* statistic. The SPSS MANOVA program, for example, provides four such statistics: Pillai's trace, Wilks's lambda, Hotelling's trace, and Roy's largest root. According to Stevens (2002, pp. 243–244), there is evidence that in practice these statistics tend to produce similar outcomes, and in general Wilks's lambda is as good a choice as any. The results in the table show that only the main effect for professional relations is statistically significant against an alpha of .05. Since Wilks's lambda indexes unexplained variance, it appears that professional relations explains about 9% of the variance in the composite variable of mental and general health. As in ANOVA, we could pursue a multivariate contrast analysis to locate the specific differences among the three relations groups. However, since we have already explored the principles of this technique in sufficient detail for our limited purposes, we will instead focus on the task of tracking down which dependent variable(s) are affected by professional relations. In other words, having formed the composite to undertake multivariate testing, we now want to unpack it again to locate the destination of the demonstrated main effect.

This unpacking can be accomplished in various ways, each with its costs and benefits. Conceptually the easiest strategy is to conduct a separate ANOVA for each of the dependent variables. When the effects of professional relations on mental health and general health are examined with two separate one-way ANOVAs, the resulting F statistics are 5.94 ($p = .003$) and 4.24 ($p = .016$), respectively. From this it appears that professional relations impacts on both health variables. Subsequent post hoc Tukey comparisons show that whereas there are mental health differences between the poor and moderate relations groups and between the poor and good relations groups, in the case of general health, only the poor and good relations groups differ. The benefits of being able to track down where professional relations have impact in this way are clear, but there is a cost. This is that we have now lost statistical control of the relationship between the two dependent variables. The two ANOVAs are linked by their dependent variables and may thereby be confounded.

There are various ways to disentangle effects on multiple dependent variables while maintaining statistical control, two of which are commonly used. The first is to examine the discriminant coefficients used in the composite variable for the effect of interest. As we noted earlier, these are calculated in a way that controls for any correlations among the dependent variables. Moreover, if they are calculated in standardized form, they can be compared with each other to gauge the relative importance of each dependent variable. In the present analysis the standardized coefficients for the professional relations main effect are .719 for mental health and .471 for general health. These figures suggest that the effect of relations on the composite variable is located more in the mental health than in the general health variable. However, although this analysis has maintained statistical control of the dependent variables' association, the conclusion is rather indirect and is not in itself subject to statistical testing.

An alternative strategy can be found in **stepdown analysis.** This is akin to hierarchical regression in which variables are added into an analysis one by one. A variable is tested as it enters in such a way that adjustments are made for its correlations with other variables entered earlier in the sequence. For the present analysis of the main effect of professional relations, the first step is a one-way ANOVA with mental health as the dependent variable. As we know, this provides an *F* statistic of 5.94 ($p = .003$). At the second step the general health variable is entered, and its relationship with professional relations is assessed while controlling for the presence of the mental health variable. This adjusted or stepdown *F* is 1.64 ($p = .197$). From this it appears that the effect of professional relations is confined to mental health; general health has no effect when its association with mental health is controlled. However, it should be borne in mind that the results of a stepdown analysis are contingent on the *order* in which the dependent variables are entered. If we enter general health first, its *F* becomes 4.24 ($p = .016$), as in the uncontrolled ANOVA above. The stepdown *F* for mental health when it is entered second is 3.28 ($p = .04$). So this analysis suggests that professional relations influences *both* types of health, as suggested by the separate ANOVAs. Moreover, either conclusion could be considered as consistent with the pattern of the coefficients. The message then is that while stepdown analysis is a powerful tool, it should not be used unless the analyst has a strong justification for the order in which variables are entered.

The use of MANOVA confronts the analyst with a plethora of decisions to be made. How should the sums of squares for the main and interaction effects be calculated to exercise appropriate statistical control? What multivariate statistics should be used to test their effects on the set of dependent variables? What strategies should be used to locate effects for particular dependent variables? What contrast analyses should be done to locate the effects within the levels of the independent variables? In these multiple analyses of the independent and dependent variables, what adjustments need to be made to combat the inflating alpha problem? These and other issues are complex, and accordingly, we have done no more than rapidly traverse the area to gain some orientation. Guides to more detailed discussions are provided at the end of the chapter. Meanwhile, it may be helpful to see some of the main ideas in action by revisiting a study that made use of MANOVA.

In Chapter 5 we discussed a study of adult attachment styles conducted by Diehl et al. (1998). Among other analyses, this study used MANOVA to examine the association of attachment style and age with 20 personality variables such as well-being, sociability, and self-acceptance. This analysis treated attachment style (secure, dismissing, preoccupied, and fearful) and age (young, middle-aged, and older adults) as factors, and the 20 personality attributes as dependent variables. Accordingly, the analysis can be described as a 4×3

factorial MANOVA with two possible main effects and one possible interaction effect. This MANOVA produced no main effect for age and no interaction effect, so attention focused on the main effect for attachment style. The multivariate test for this effect produced a Wilks's lambda of .71 with an F of 1.67, $p < .01$. Thus attachment style appeared to share nearly 30% of its variance with the set of personality variables. The multivariate effect was followed up by separate ANOVAs of each of the 20 dependent variables. These analyses showed statistically significant differences ($p < .001$) in attachment style for 9 of these personality attributes: dominance, capacity for status, sociability, social presence, empathy, good impression, communality, well-being, and achievement through conformance. Post hoc comparisons using Tukey's honestly significant difference test were then conducted to identify which particular attachment styles were related to these personality variables. The pattern of results was complex, but showed, for example, that respondents with a secure attachment style had higher mean scores on well-being and on achievement through conformance than did those with a fearful or preoccupied attachment style. This provides a good illustration of how MANOVA can be used to dissect a very complex data set and to elicit clear messages from it.

6.4 WITHIN-SUBJECTS ANOVA AND MANOVA

In all of the ANOVA and MANOVA examples we have discussed so far, the different categories in each factor have contained different subsets of cases. For example, each case in the preceding study by Diehl et al. appeared in one, and only one, attachment-style group. A factor that has this feature is referred to as "**between-subjects**" because any comparisons across categories will be between different subjects or cases. Many studies, however, include factors that do not have this feature. If the categories of a factor do not contain different cases, the factor is referred to as "**within-subjects**." The most common reason for this is that the same cases appear in more than one category, a so-called **repeated-measures** factor. In this situation cases are being compared with themselves—the comparisons are within-subjects. Imagine an experiment that measures well-being before and after physical exercise to see if the latter has any beneficial effect. Here the two-category factor would be time (before and after), and the dependent variable would be the well-being score. The analysis would consist of a comparison of the before and after means using

ANOVA or a t test. However, the fact that time is a repeated measures factor, where the same cases generate both means, makes the ANOVA technique we have discussed so far an inappropriate choice.

Remember that an ANOVA consists essentially of a comparison of group differences with individual differences. The former are indexed with a between-groups sum of squares and variance, and the latter are captured with a within-groups sum of squares and variance. When the groups contain the same people, the problem lies in how we calculate individual differences. Usually this is achieved by calculating how much individuals differ from their particular group mean, and then summing these differences across the groups. When the groups contain different individuals this is a legitimate addition of independent packages of variability. However, if the groups contain the same individuals, the sum will be artificially inflated to the extent that they behave consistently under the different conditions. In the exercise example individuals' before and after well-being scores will be correlated simply because they are provided by the same people. So our estimate of individual differences, pooled across the two conditions, is likely to be an overestimate as we combine the overlapping components.

To counter this inflationary problem, the within-groups sum of squares is adjusted so that any variability due to individuals' consistency across conditions is removed. What is left, the residual or error sum of squares, is then converted into a variance or mean square as usual. This deflated variance then becomes the denominator for the F statistic that tests the effect for the within-subjects factor. The numerator is the between-groups variance calculated in the usual way. In summary then, the key concern when analyzing a repeated measures factor is to calculate the F statistic using the residual variance rather than the within-groups variance in order to remove the bias introduced by comparing cases with themselves. This adjustment is needed for the analysis of any within-subjects main effect, or for any interaction that includes this type of factor. It is also required regardless of whether the overall analysis is an ANOVA or a MANOVA.

To see all of these ideas in action, we can turn to a real experiment on exercise and well-being conducted by Jerome et al. (2002). Their particular interest was in how feelings of self-efficacy with respect to physical exercise influence emotional reactions to exercise. Participants were randomly assigned to a low- or high-efficacy group in which they were given false feedback about their capacity for physical exercise. The low-efficacy group received detailed information that painted a dismal picture, while the high-efficacy group received a much rosier picture. Participants subsequently engaged in structured exercise activities and completed measures of emotional reactions before and after the session. This is a very crude summary of a complex and sophisticated study, but it is sufficient for present purposes. The design therefore consisted of one between-subjects factor (low vs. high efficacy), one repeated measures factor (before and after), and multiple emotional response measures as dependent variables. So from a data analytic perspective, the overall design would be labeled a 2×2 mixed model (i.e., with both a between- and a within-subjects factor) MANOVA.

One of the reported MANOVAs was conducted on a set of three emotional responses: well-being, distress, and fatigue. The multivariate tests of the two main effects and the interaction effect showed that only the last was statistically significant with an F of 6.2, $p < .01$. This multivariate interaction effect was explored by analyzing each of the three dependent variables with separate ANOVAs. These showed that the interaction effect applied to all three variables. Participants in the low-efficacy group reported decreased well-being and increased distress and fatigue between the before and after measurements, while those in the high-efficacy group reported no appreciable changes. Note that no subsequent contrast analyses needed to be undertaken since both factors had only two categories. In general, this study just provides another illustration of MANOVA and ANOVA strategies we discussed in previous sections. The key new point, though, is that all of the calculations involving the time factor have been adjusted to take account of its within-subjects status.

As long as a within-subjects factor has no more than two categories, the procedures outlined above for ANOVA are legitimate. However, an added complication arises when a within-subjects factor has more than two categories. Imagine that we want to find out whether listening to classical music reduces anxiety immediately and for some time after. Anxiety measurements are obtained from a group of participants before they listen to a piece of music, immediately afterward, and for a final time one hour later. This imaginary study has one within-subjects factor—time, with three categories, and one dependent variable—anxiety. The three mean anxiety scores could then be analyzed with a one-way within-subjects ANOVA. However, to be legitimate, such an analysis would need to satisfy an assumption called the sphericity assumption. We will defer an explanation of this assumption until the next section. For now, it is only necessary to appreciate that the assumption is easily violated and that the consequences of this violation can be serious.

There are various ways to deal with this problem, but the strategy of most interest in the present context is to magically convert the within-subjects ANOVA into a MANOVA. As we just noted, in the former design we have a three-level factor and a dependent variable. So the three means are viewed as mean scores on the dependent variable, one for each level of the factor. The conversion begins by transforming the three means into a set of nonoverlapping mean differences. Testing the equality of three means is equivalent to testing simultaneously that the first is equal to the second and that the second is equal to the third. In general, a set of means can be transformed without loss of information into a set of mean differences that is one less than the number of means. So our three mean scores become two mean difference scores. The second step in the conversion is to treat these difference scores as two dependent variables in a MANOVA. This feels like an alarming move as it seems to leave the independent variable box empty. Again, this demonstrates the difference between the researcher's need to assign independent and dependent variables and the blindness of statistical techniques to such a distinction. Whatever the transformation, the fact remains that we are still analyzing the differences among three means. But the gain is that unlike within-subjects ANOVA, MANOVA does not require the sphericity assumption. It is therefore generally recommended that any designs involving at least one within-subjects factor with more than two categories be analyzed with MANOVA.

This strategy can be seen in a real study of music and anxiety conducted by White (1999). In this study, 45 patients who had suffered a recent heart attack were randomly assigned to one of three groups. The experimental group spent 20 minutes listening quietly to classical music; the second group relaxed for 20 minutes without music; and the third control group spent the 20 minutes engaged in their usual activities. Measures of anxiety and of five physiological processes were taken just before the 20-minute period, just after, and one hour later. From one perspective this study has one between-subjects factor (the three groups), one within-subjects factor (three time periods), and six dependent variables. This would make it 3 × 3 MANOVA design with six dependent variables. However, recognizing the potentially problematic nature of the within-subjects time factor, White converted it into two dependent difference variables in the manner described above. This strategy is called a **doubly multivariate design** in which the multiple "true" dependent variables are joined by the "quasi" dependent variables that represent the within-subjects factor.

This monster MANOVA uncovered a multivariate group by time interaction effect. To discover to which variables the interaction applied, a separate MANOVA was conducted for each of the six dependent variables. These were MANOVAs rather than ANOVAs because the within-subjects time factor was implicated and continued to need the protection afforded by MANOVA. The MANOVA for anxiety produced a statistically significant interaction. Subsequently, post hoc comparisons using Tukey's honestly significant difference statistic, showed that reductions in anxiety immediately after the 20-minute period and one hour later were greater in the music group than in the other two groups. Again this study illustrates well the use of MANOVA to exercise multivariate control over complex patterns of relationships and then systematically to isolate specific effects of interest.

6.5 ISSUES OF TRUSTWORTHINESS IN MANOVA

Most of the issues in this short section will be familiar from discussions in earlier chapters, though some of the details are specific to MANOVA. As usual, the issues revolve mainly around sampling, measurement, and statistical assumptions. With respect to sample size, a minimum requirement is that each cell of the design has more cases than there are dependent variables. Beyond this minimum, the analyst faces the usual issues of obtaining enough cases to achieve appropriate levels of Type I and II error given the expected size of effect, but also given the number of cells or groups, and the number of dependent variables. Stevens (2002, pp. 626–629) has provided a useful table for the calculation of MANOVA sample sizes as a function of all of these. The table shows, for example, that if an analyst required an alpha of .05 and power of .80 to detect a moderate effect size with a MANOVA consisting of four groups and three dependent variables, at least 58 cases would be needed in each group. It is not a requirement that all group sample sizes are equal, but if they are not, care is needed to calculate sums of squares in a way that adjusts for an unbalanced design.

The independent variables in a MANOVA are usually categorical, and the dependent variables have been measured on at least an interval scale. As we will see in the next section, though, interval level independent variables may also be included, in which case the technique required is a variant called analysis of covariance. Ordinal-level independent variables may also be included by specifying contrast analyses in appropriate ways. It is assumed that all variables have been measured with adequate reliability and validity. The consequences of failing to achieve this are complex, but one of the most worrisome as usual is the tendency for unreliable measurement to attenuate effects and make them more difficult to detect.

Turning to statistical assumptions, we find that these are essentially the same as for discriminant analysis, except that the independent and dependent status of the variables is reversed. It is assumed that the dependent variable scores are multivariate normal in each category of the independent variable(s) and that their variances and covariances are equal across the categories: the assumption of homogeneity of variance-covariance matrices. As noted in Chapter 5, this latter assumption can be checked with Box's M statistic, though this overly sensitive statistic should be used with caution (Tabachnick & Fidell, 2001, p. 330). Also as in discriminant analysis, MANOVA assumes independence of cases and multivariate linearity of the dependent variables in each category. Multicollinearity (very high correlations) in the dependent variables may be problematic, as may the presence of outliers.

In Section 6.4 the point was made that if a within-subjects factor has more than two categories, a new and demanding assumption arises for ANOVA but not for MANOVA: the **sphericity assumption.** It is useful to discuss this further at this point because questions can sometimes arise about the comparative trustworthiness of results from an ANOVA and MANOVA of the same data. First, what does the sphericity assumption require? As we noted earlier, an analysis of a within-subjects factor can be seen as a comparison of a set of difference scores across pairs of categories. If this set of differences is analyzed by ANOVA, the sphericity assumption requires essentially that the variances of the difference scores are equal and that the scores on each difference are uncorrelated. Note that if there are only two categories, there is only one difference score and so the assumption cannot apply.

The sphericity assumption is easily violated and can lead to an increase in Type I error. Violation of the assumption can be detected with the **Mauchley test,** but like Box's M, the test can be overly sensitive. Why then not routinely analyze within-subjects factors with MANOVA, as described above, and avoid the sphericity assumption altogether? The main reason is that under some circumstances, the ANOVA approach is preferable because it can provide more statistical power. Moreover, the ANOVA significance tests can be adjusted with a **Greenhouse-Geisser** or **Huynh-Feldt correction** to compensate for the increased error rate. These are complex issues, and useful discussions of them can be found in Stevens (2002, pp. 500–506, 509–510) and Tabachnick and Fidell (2001, pp. 421–423). Probably the best general piece of advice for the analysis of within-subjects factors is to use both ANOVA and MANOVA, and to consider the best course of action in light of the overall pattern of results. The SPSS package, for example, encourages such an approach by routinely providing both ANOVA and MANOVA results when the latter is used to analyze a design that includes within-subjects factors.

MANOVA and ANOVA are often used for the analysis of experimental data that result from the systematic manipulation of independent variables. Since experiments are generally regarded as particularly trustworthy sources of information, it is appropriate to finish this section with a comment on MANOVA analyses of experimental data. The comment is an echo of our discussion at the end of Chapter 3 about the limitations of statistical control. However complex a MANOVA analysis may be, its results cannot compensate for a poor experimental design. A good experiment is one in which the manipulation of the independent variable(s) is conceptually appropriate and rigorously executed. The experimental conditions must be explicitly defined and controlled using appropriate strategies such as random allocation to groups, balanced sequencing of manipulations, and physical control of nuisance variables. The use of MANOVA, or any other multivariate technique, can provide limited statistical control of confounds, but this contribution is minor compared with the critical role of good experimental design and control. When we are enmeshed in detailed discussions of the complexities of MANOVA, it is easy to overestimate its role in the research process. It is important to keep in mind that ultimately it is no more than a sophisticated strategy for analyzing mean differences. Interpretation of these differences remains as problematic as ever and has to be framed more in terms of theory, measurement, and design rather than of statistics.

6.6 ANALYSIS OF COVARIANCE

As we noted earlier, both ANOVA and MANOVA can be extended to include interval-level independent variables, known in this context as covariates. An

ANOVA with at least one covariate is called an **analysis of covariance** (ANCOVA), while the multivariate version is known as a multivariate analysis of covariance (MANCOVA). Covariates are included in analyses to provide statistical control of variables that might confound the main and/or interaction effects of the factors and to enhance power by reducing individual differences on the dependent variable. We will exemplify a simple ANCOVA and MANCOVA with the nursing data, focusing as usual on the effect of professional relations on health.

The first ANCOVA analysis treats professional relations as a three-category factor (poor, moderate, and good), age in years as a covariate, and mental health as the dependent variable. So the objective of this analysis is to test the effect of professional relations on mental health, while controlling for any effect of age. Although age is included essentially as a control variable, note that we can also test its effect on mental health both on its own and interacting with professional relations. When we conducted an ANOVA without controlling for age in Section 6.1, we found an F value of 6.98, $p = .001$. The present ANCOVA produces an F for professional relations of 6.34, $p = .002$. This indicates that age is not acting as a confound, since it makes very little difference whether it is controlled or not. Moreover, the ANCOVA shows that age has no effect on mental health either in its own right or by moderating the effect of professional relations.

To illustrate the use of MANCOVA, we can repeat the ANCOVA but now include general health as a second dependent variable. The multivariate test of the professional relations effect provides a Wilks's lambda of .912, with an F of 3.69, $p = .006$. This shows that professional relations has an effect on the two dependent variables considered as a set, even when age is statistically controlled. Again, there are no statistically significant effects involving the age covariate. This outcome then leads, as we saw in earlier sections, to the usual efforts to locate which dependent variables are affected and which group differences account for the effect. So, in general, the shift to ANCOVA and MANCOVA does not alter the overall data analytic strategy.

The inclusion of covariates does, however, introduce two further assumptions to those found in ANOVA and MANOVA. The first assumption is that there is a linear relationship between the dependent variable and any covariates. To understand the second, it is helpful to remember that an ANCOVA is essentially a regression analysis under another name. This means that the relationship between a covariate and the dependent variable will be described by a regression slope. More precisely, there will be a regression slope describing this relationship for each category of the factor. The second new assumption is that these regression slopes are equal across the categories. For example, in the nurses' ANCOVA it is assumed that the regression slopes showing the relationship between age and mental health are the same for each of the three professional relations subgroups or categories. Assessing this assumption becomes complicated as the number of factors and covariates increases. An excellent discussion of this and other issues in the analysis of covariance can be found in Chapter 9 of Stevens's (2002) text.

6.7 FURTHER READING

Keppel, Saufley, and Tokunaga (1993) provide an excellent basic introduction to ANOVA, which is elaborated further in Keppel (1991). The topic of contrast analysis is also explored fully in Keppel (1991) and in Rosenthal and Rosnow (1985). Discussions of interactions in factorial ANOVA can be found in these sources and in Jaccard (1998). Stevens (2002) gives extensive coverage of all aspects of ANOVA and MANOVA, including chapters on repeated measures analysis and analysis of covariance. Tabachnick and Fidell (2001) similarly devote multiple chapters to all of these techniques.

SEVEN

FACTOR ANALYSIS

The fundamental aim of any multivariate analysis is to detect coherent patterns in complex data. Although the sought-after patterns are most often associations between independent and dependent variables, this is not always the case. In this chapter we explore a family of techniques called factor analysis, which is used to detect patterns in a set of interval-level variables, all of which are treated as if they were dependent. The goal is to see if it is possible to reduce the set of measured variables to a smaller set of underlying factors. Thus the factors are seen as "latent" independent variables, which are in some sense responsible for the observed dependent variables. The starting point for a factor analysis is the network of pairwise relationships among the dependent variables. In terms of statistical accounting, the objective is to find a set of factors that account for this pattern of associations. As ever, this search is guided by a principle of coherent parsimony, meaning that the data should be accounted for with the smallest number of interpretable factors without any serious loss of information.

Broadly speaking, there are three types of situations in which an analyst might want to undertake such an exercise in data reduction. In the first type the emphasis is on *replacing* a set of variables with a smaller set of factors. As we have seen, despite the capabilities of multivariate analysis, it is often desirable to reduce the number of independent or dependent variables in an analysis to render it more powerful and manageable. Combining variables by factor analysis is one way to achieve this. Moreover, as will become apparent, it is possible to generate factors that are uncorrelated, so using such factors may also help to combat problems of multicollinearity. The second type of situation focuses on the characteristics of one or more measurement instruments. If, for example, the analyst is using a questionnaire measure of well-being that is believed to assess three different facets of the construct, factor analysis can be used to examine whether the item responses cohere appropriately into three clusters. In the third type of situation the emphasis is more theoretical in that underlying factors are seen as *explaining* the patterning in the data. These variable reduction, psychometric, and theoretical uses of factor analysis can be differentiated, but in practice they will often overlap and so should not be seen as mutually exclusive.

In Section 7.1 we gain an overview of factor analysis, focusing as usual on how it relies on composite variables to achieve its objectives. Section 7.2 shows factor analysis in action with details on how an analysis proceeds and on the statistics that guide the analyst. The first two sections again make use of the nurses' data but finish with another example from the well-being research literature. Section 7.3 discusses the issues of trustworthiness that confront the factor analyst. Finally, in Section 7.4 we introduce an approach called **confirmatory factor analysis.** All of the analyses in the first three sections are exploratory in that no formal hypotheses are tested about the nature of the factors that might be detected. The more advanced technique of confirmatory factor analysis allows for such formal tests, for example, about the number of factors that might be present or about their contents. Also, as we have seen in other contexts, it allows for the formal comparison of different models or factor solutions. Confirmatory factor analysis is one of a group of advanced techniques called structural equation models. Since these "second-tier" techniques lie beyond the scope of a brief introductory text, we do little more than outline the logic of confirmatory factor analysis and show it in action. That said, the fundamental approach should be easily comprehended since it shares much of the same framework as exploratory factor analysis.

7.1 THE COMPOSITE VARIABLE IN FACTOR ANALYSIS

Most of our analyses of the nurses' data have revolved around their mental health, as measured by the mental health subscale of the SF-36 (Ware & Sherbourne, 1992). The full SF-36 health measure comprises eight subscales, all of which were completed by the sample of nurses. We will now use all of the SF-36 data to explore the fundamentals of factor analysis. The subscales are mental health (MH), which assesses psychological distress and well-being; role emotional (RE), which assesses activity limitations due to emotional problems; energy/vitality (EV), which assesses energy and fatigue; social function (SF), which assesses social limitations due to physical or emotional problems; physical pain (P); general health perception (GHP); physical function (PF), which assesses limitations in physical activities due to health problems; and role physical (RP), which assesses limitations in usual activities because of physical health problems. Scores on all of these subscales are scaled to a possible minimum of zero and a possible maximum of 100, where a higher score indicates better health.

After some loss of cases due to missing data, scores on all of the SF-36 subscales were calculated for 151 nurses. The simple correlations among all pairs of scores are shown in Table 7.1 and form the starting point for our discussion of factor analysis. Note that this focus means that all that follows is essentially an analysis of *individual* differences.

Table 7.1 Simple Correlations Among SF-36 Subscale Scores

	MH	*RE*	*EV*	*SF*	*P*	*GHP*	*PF*	*RP*
Mental health (MH)		69	.76	.69	.25	.44	.24	.36
Role emotional (RE)			.62	.66	.20	.39	.17	.42
Energy/vitality (EV)				.73	.42	.62	.20	.47
Social function (SF)					.40	.51	.28	.51
Pain (P)						.44	.37	.33
General health (GHP)							.32	.51
Physical function (PF)								.30
Role physical (RP)								

All of the subscales are positively correlated, and all of these correlations are statistically significant at $p < .05$. But is there some simpler structure hidden beneath these 28 correlations? For example, do the first four subscales that are predominantly about mental health form a subgroup separate from the remaining subgroup of four subscales that are predominantly about physical health? In other words, does the pattern of correlations reflect two underlying factors? If so, this may have implications for ways in which the subscales might be combined in other analyses, for the psychometric properties of the subscales, and for theories about the structure of self-reported health. Whatever our motivation for the factor hunt, we need some statistical tools to undertake it. Even though we have only eight variables, it is not easy to tell by inspection of Table 7.1 whether the correlations within the mental and physical health domains are higher than those across them. Moreover, these correlations provide only a bivariate perspective on the network of relationships; as usual a multivariate view would be more powerful.

The first step in a factor analysis is called the extraction phase. Imagine that all of the individual differences on all of the variables are thrown together in a large box. The extraction process involves dipping into the box and trying to remove the largest possible package of differences. This package is the first factor to be extracted. Then comes another dip into the box to remove the next biggest package, with the constraint that the information contained in each package is independent of any other. This process of extracting mutually exclusive factors, decreasing in magnitude, can continue in principle until there are as many factors as variables. Since this is hardly parsimonious, the extraction process is usually halted well before that. The extraction process raises a number of questions that must be answered by the analyst, notably: Should all of the individual differences be placed in the box? Which of a variety of extraction techniques should be used? And how many factors should be extracted? But we will postpone an exploration of these questions until the next section.

A more immediate question is: What exactly are these packages called factors? It will come as no surprise to learn that they are composite variables. In the present example, where there are 8 variables, a **factor** can be expressed as:

factor = [coefficient 1(DV1) + coefficient 2(DV2) + ...
+ coefficient 8(DV8)]

This says that a factor is a weighted composite of all of the variables. Strictly speaking, the variables are neither independent nor dependent, but it is helpful to think of them as dependent on the underlying factors. Note that since all of this is in standardized form, there is no constant. There will be as many factors as there are variables, and each factor will have a different set of coefficients. The coefficients are chosen so that they maximize the amount of individual differences or variability captured by "their" factor, while ensuring that this variability is uncorrelated with that contained in any other factor. The actual amount of variance contained in a factor is indicated by its **eigenvalue.** So, as in the case of discriminant analysis, another way to express the rule for generating factors is in terms of maximizing eigenvalues.

Having extracted an appropriate number of factors less than the number of variables, the analyst now needs to be reassured that each is interpretable and that jointly they account for an acceptable portion of the individual differences, that is, provide a good fit to the data. Or to put the latter point in reverse, we need reassurance that the chosen factors have not lost too much of the information in the original data. The coefficients used to extract factors are not easily interpreted. Instead analysts focus on various other types of coefficients that quantify the *correlations* between variables and factors. Most attention is usually paid to factor loadings: the correlations between variables and factors adjusted for any correlations among the factors. Even the factor loadings that emerge from the initial extraction are rarely interpretable as they stand, for reasons we will discuss later. Accordingly, the interpretation phase is usually preceded by a transformation of the loadings called factor rotation. This process in turn leads to yet more choices: this time among a variety of rotation techniques. Again we will postpone details on rotation, interpretation of loadings and coefficients, and assessing the fit of a particular factor solution until the next section.

This brief, and probably dizzying, overview should at least make a few initial points clear. Factor analysis is a complex, multistage technique that requires the analyst to make a variety of decisions. It follows from this that a sensible analyst will explore the consequences of making different decisions by comparing solutions with different numbers of factors extracted and rotated in different ways. The technique then is inherently comparative in its approach to data analysis. It further follows that there is no such thing as a definitive factor solution, although some confidence is warranted if different strategies produce very similar solutions, which fortunately happens more often than not. We now turn to an actual analysis of the SF-36 data to make all of these ideas more concrete.

7.2 FACTOR ANALYSIS IN ACTION

Before we embark on a factor analysis, it is prudent to ensure that the variables in the analysis are sufficiently interconnected to make them "factorable." If the variables have a weak connecting structure, there is little point in trying to characterize it in the form of underlying factors. The magnitude and statistical significance of the simple correlations provide some guide to structure but, as we noted earlier, a multivariate perspective would be more helpful. A useful statistic for this purpose is the partial correlation, which we first encountered in Chapter 4. A **partial correlation** is the correlation between any two variables in a set when all other variables are statistically controlled. If the partial correlation is notably smaller than the simple correlation between two variables, this suggests that one or both of the variables is strongly linked to the set of variables that are being controlled. The bigger the difference between the simple and partial correlations, and the smaller the partial correlation, the more evidence there is that the correlated variables are tied into a multivariate structure. For example, the simple correlation between mental health and general health perception is .44, while their partial correlation drops to –.06. The strong effect of partialing out their relationships with the other six variables suggests that one or both of the two variables is embedded in a multivariate structure.

It is helpful to examine the partial correlations among all pairs of variables, usually known as **anti-image correlations** in the context of factor analysis. (Strictly speaking, an anti-image correlation is a partial correlation with a reversed sign.) Further insights into factorability can be gained by using the **Kaiser-Meyer-Olkin (KMO) measure of sampling adequacy.** The KMO statistic summarizes the partial correlations for a particular variable and compares them with that variable's simple correlations. This comparison is expressed as an index with values between zero and one. A value less than .5 is regarded as indicating a structure that is unacceptable for factor analysis. In the present data the KMO statistics for all variables are in the range .8 to .9. Since the authors of the statistic regard .8 as "meritorious" and .9 as "marvelous" (Hair, Anderson, Tatham & Black, 1998, p. 99), we can rest assured that there is a strong multivariate structure available for analysis. These individual variable KMO values can also be averaged to provide a value for the whole set of variables: a meritorious .85 in this case. Thus the partial and simple correlations provide the means to examine the relational structure of the data at the level of the variable, pairs of variables, and the total set of variables. With these insights, we can now move to the factor analysis proper.

7.2.1 Extracting and Rotating Factors

The extraction of factors can be achieved in a variety of ways. Fundamental to our choice is the decision on whether we want to analyze all of the individual differences in the data or just a selected portion. To clarify the nature of this decision, we can focus for a moment on the mental health and general health perception variables. If we pool all of the individual differences or variability for these two variables, we can identify three types of variability. The first type is shared or common variability, as indexed by the correlation between mental health and general health perception. Next there is the unshared or unique variability for each of the two variables. These are the individual differences in mental health and general health perception that cannot be accounted for by the association between them; they are unique to each variable. Finally, within the unique variability for each variable there are differences due to random measurement error, that is, unreliability.

Broadly speaking, the analyst can choose whether to extract factors from all of the individual differences or only from those reflecting shared variability. The former choice leads to an extraction technique called **principal components analysis (PCA).** The latter choice leads to a family of extraction techniques called **factor analysis (FA).** So, confusingly, the term factor analysis is used in two ways: as a loose generic term for all extraction techniques including PCA and as a label for any technique that extracts factors from shared variability only. Later we will discuss the issues that may influence the PCA versus FA decision. For now, in order to achieve a rapid overview of the analytic process, and in the spirit of exploratory factor analysis, we will use PCA and a parallel technique called **principal axis factor analysis (PAFA).** They are parallel in that both extract uncorrelated factors in a sequence whereby each successive factor accounts for a decreasing amount of variance. Their main difference, as we noted, is that PCA extracts factors (or, strictly speaking, components) from all available variance, while PAFA extracts factors only from shared variance.

The production of a series of uncorrelated factors that account for decreasing amounts of variance is a tribute to an elegant statistical process, but it will not necessarily generate factors that are interpretable. As we will soon see, the interpretation process focuses on how individual variables contribute to a factor. The main tool used is the **factor loading,** which is the correlation between a variable and a factor. So the higher its loading, the more a variable is seen as contributing to, and therefore defining, a particular factor. The loadings produced by PCA and PAFA may be misleading because of the extraction requirements for uncorrelated factors that account for decreasing amounts of variance. The next phase of the analytic procedure offers the opportunity to relax one or both of these requirements using a technique called **factor rotation.**

In essence, rotation transforms the values of the factor loadings to meet the requirements specified by the analyst. The first decision is whether to maintain uncorrelated factors or to allow them to be correlated. Rotation techniques that achieve the first are called **orthogonal rotations;** those that achieve the second are called **oblique rotations.** Again, we will postpone any discussion of the relative merits of these strategies until we have completed an overview of the whole process. Also again, we will proceed with both types of rotation to explore the outcomes. The second rotation decision concerns which particular orthogonal and/or oblique rotation technique should be adopted, since both types can be achieved in various ways. A discussion of specific rotation techniques is beyond the scope of this account and can be pursued in the readings provided at the end of the chapter. We will proceed with two popular techniques: an orthogonal strategy called **varimax rotation** and an oblique strategy called **direct oblimin** rotation. Varimax rotation aims to maximize the spread of variance across the extracted factors, that is, to remove the constraint that each succeeding factor should account for less variance. In loose terms, oblimin rotation aims to minimize the correlations of loadings across factors. Both of these strategies, and others, are designed to achieve what is called **simple structure.** Crudely put, this means achieving a factor solution in which each factor is well defined by a subset of variables, and each variable has a clear "home" factor.

The final decision required for the extraction and rotation phases is the number of factors that should be analyzed. Earlier we suggested that two factors might underpin the eight SF-36 variables—a mental health factor and a physical health factor. Exploratory factor analysis may begin from such a proposal or might allow the data to suggest the number of underlying factors. For now we will constrain the analysis to two factors and return to the topic of data-driven exploration later.

7.2.2 Interpreting Factors

With this rapid overview in place, we can now try our hand at evaluating some actual results. This evaluation revolves around two questions: What do the factors mean? and How well do they represent the data? We will explore the first question in this subsection and the second in the following subsection. Table 7.2 shows the rotated factor loadings for four analyses: a PCA and PAFA extraction of two factors, each with a varimax (orthogonal) and oblimin (oblique)

Table 7.2 Rotated Factor Loadings for Principal Components and Principal Axis Analyses

Extraction:	*Principal Components*				*Principal Axis*			
Rotation:	*Orthogonal*		*Oblique*		*Orthogonal*		*Oblique*	
Factor (F):	*F1*	*F2*	*F1*	*F2*	*F1*	*F2*	*F1*	*F2*
Mental health	**.87**	.12	**.91**	.08	**.84**	.20	**.91**	.07
Role emotional	**.86**	.06	**.91**	.15	**.78**	.16	**.84**	.09
Energy/vitality	**.84**	.31	**.84**	.12	**.77**	**.42**	**.74**	.21
Social function	**.81**	.33	**.80**	.15	**.73**	**.42**	**.70**	.22
Physical function	.02	**.77**	.13	**.81**	.10	**.48**	.05	**.52**
Pain	.17	**.77**	.03	**.77**	.15	**.64**	.04	**.68**
General health	**.49**	**.60**	.41	**.52**	.39	**.63**	.23	**.59**
Role physical	**.46**	**.52**	.39	**.44**	.37	**.51**	.26	**.45**

rotation. As we noted, the loadings show the associations between each variable and each factor. For the orthogonal rotations, the loadings are just simple correlations; for the oblique rotations, the loadings are regression coefficients that adjust for any correlation between the factors.

In order to interpret the loadings, it is necessary first to decide how large a value is required for a loading to be reliable and noteworthy. A practice has developed of using .30 as a minimum loading mainly on the grounds that such a variable would be accounting for nearly 10% ($.3^2$) of the variance in a factor. But this has been criticized as ignoring the effect of sample size. Stevens (2002, pp. 393–394) has reviewed these arguments and provided a useful table that allows the analyst to determine a minimum value reflecting both stringent statistical significance ($p < .01$) and size of contribution. On this basis, a loading would need to be at least .42 to be worthy of interpretation in the present example; all loadings that meet this requirement are shown in bold in Table 7.2.

Overall, the analyses suggest that Factor 1 represents mental health and Factor 2 represents physical health. However, the pattern of loadings differs across the four solutions in interesting and instructive ways. The oblique solutions both show a separation of the loadings into two factors, whereas the orthogonal solutions do not. The PCA orthogonal loadings show overlap for the general health and role physical variables; and the PAFA orthogonal loadings show overlap for the energy/vitality and social function variables.

(Note, though, that the latter loadings are barely statistically significant.) For the oblique solutions, it is possible to calculate the correlations between the factors, which are .40 for the PCA and .55 for the PAFA. This outcome and the fact that the subscales are united by a self-report health construct suggest that an oblique rotation is preferable. Note that an orthogonal rotation *forces* the factors to be uncorrelated, whereas an oblique rotation only produces correlated factors *if* they emerge from the solution. So, an oblique rotation may produce uncorrelated factors, but an orthogonal rotation can only produce uncorrelated factors. In the present case an orthogonal rotation seems inappropriate and results in distortions in the factors. Note, though, that there are occasions when an analyst actually requires uncorrelated factors, and then an orthogonal rotation would be mandatory.

If we compare the two oblique solutions, it is notable that the PAFA extraction achieves a cleaner separation of the factors than does the PCA. This might be expected if we reflect on the difference between the two extraction techniques. Remember that the PCA is extracting factors from all of the available variance in the eight variables: shared, unique, and error variance. The presence of the last of these in particular may well make it harder to clearly expose the underlying structure. In contrast, the PAFA has extracted the factors only from shared variance. This strategy runs the risk of losing important information about a variable but has the advantage of cleaning out the error variance before the extraction begins. In the present case the Cronbach's alphas for the eight subscales suggest that measurement error is not a problem of any magnitude. It may be, therefore, that the difference in the factor separation between the PCA and PAFA solutions is more due to the presence of unique variance in the former. The extent to which this is a problem depends on the goals of the analysis. If the objective were to account for the relationships among the subscales with some underlying structure, then the PAFA route is clearly preferable. But if the objective were to replace the eight subscales with two super subscales, then retaining unique variance through PCA may well be preferable.

The factor loadings are a useful interpretive tool, but other tools are available. To discuss these, we will narrow our focus to the oblique solutions for the PAFA extraction. As we noted earlier, the loadings in an oblique solution are adjusted for any correlation among the factors. It may also be instructive to look at the simple correlations between the variables and factors, ignoring any correlation among the factors. The adjusted loadings are called **pattern**

coefficients, while the unadjusted loadings are called **structure coefficients.** Once an acceptable solution has been found, it is possible to calculate **factor scores.** So, if we decided to accept a two-factor solution for our health variables, we could compute two factor scores for each case: a mental health score and a physical health score. A factor score would be computed for each case using the usual composite variable strategy. For each case the scores on each of the eight variables (in standardized form) would be multiplied by a factor score coefficient, and the products summed. So each factor has a set of **factor score coefficient**s that can be used to produce factor scores. These coefficients have additional interpretive value since, unlike pattern and structure coefficients, they show the contribution of each variable to a factor while controlling for correlations among the *variables.* Jointly, the pattern, structure, and factor score coefficients provide a comprehensive view of variable-factor relationships.

These three types of coefficients are shown in Table 7.3 for the oblique PAFA solution. All three sets of coefficients for Factor 1 suggest that this mental health factor is well defined. Whether the correlations among the factors (pattern coefficients) or among the variables (factor score coefficients) are controlled or not (structure coefficients), the first four variables listed have consistently high coefficients relative to the other variables. So from both an interpretive and psychometric viewpoint, the factor is easy to defend. In contrast, the second factor presents a messier picture. The structure coefficients lose the separation between the two subsets of variables that appears in the pattern coefficients. This reflects the .55 correlation between the factors that is controlled in the pattern coefficients but not in the structure coefficients. The factor score coefficients suggest that it would not be wise to form a Factor 2 score on the understanding that it would represent physical health. The general health (.31) and pain (.29) variables would carry most weight, but then there would be a mixture of physical and mental health variables, all weighted with coefficients of .17 or .18. This demonstrates well the point that factor scores do not necessarily reproduce the profile of variables suggested by the loadings since the two sets of coefficients are calculated under different statistical constraints. Moreover, it is worth noting that factor scores in factor analysis (but not in PCA) have to be estimated and have no unique solution.

The conclusions that can and should be drawn from this sort of interpretive analysis depend, of course, on the analyst's objectives. If the intention were to explore the possible two-factor structure of self-reported health, then

Table 7.3 Pattern, Structure, and Factor Score Coefficients for a Principal Axis Factor Analysis With Oblique Rotation

Coefficients:	*Pattern*		*Structure*		*Factor Score*	
Factor (F):	*F1*	*F2*	*F1*	*F2*	*F1*	*F2*
Mental health	.91	.07	.86	.42	.32	−.10
Role emotional	.84	.09	.79	.38	.22	−.07
Energy/vitality	.74	.21	.85	.62	.28	.17
Social function	.70	.22	.82	.61	.23	.18
Physical function	.05	.52	.24	.49	−.01	.18
Pain	.04	.68	.33	.66	−.03	.29
General health	.23	.59	.56	.72	.04	.31
Role physical	.26	.45	.50	.59	.04	.17

the pattern coefficients from the oblique PAFA provide a clear picture. The structure coefficients from that analysis could be seen as appropriate, if the correlation between the factors were acceptable. Moreover, the different pictures provided by the pattern and structure coefficients could be seen as informative about the ways in which mental and physical health interconnect. In contrast, if the intention were to develop two mental and physical scales that incorporated the eight subscales, the message is that this would not be justified by the present results, at least for the physical scale. It is important to appreciate, though, that interpretation is only one criterion, albeit a crucial one, in deciding on the acceptability of a factor analysis. We now turn to another criterion—the question of how well the solutions capture the data.

7.2.3 Assessing Goodness of Fit

There are various ways to approach the question of fit. If PCA is being used, and therefore all variance is available for extraction, an obvious question is: How much of that total variance does a particular solution account for? Computerized factor analyses routinely provide this sort of information. For our PCA analyses, the first factor extracted 52.5% of variance and the second a further 14.6%, producing a total of 67.1%. After varimax rotation, this variance was redistributed 41.7% to Factor 1 and 25.4% to Factor 2, but note that this is just a redistribution, so the total stays the same. When an oblique rotation uncovers correlated factors, the redistributed variances are overlapping and so cannot be assigned clearly to factors after rotation. Whatever the rotation, it is hard to avoid asking the question of whether the total amount of variance extracted is adequate. There is no straightforward answer to this question, just as there is no easy answer to the question of how large an R^2 should be in multiple regression. If the objective is to find out whether a predetermined number of factors can be found, then the issue of total variance explained is of secondary interest anyway, as in the present case. Even if the objective is to maximize the extracted variance, it still will not be clear what types of variance—shared, unique, and error—have ended up in the solution. In general, the amount of variance extracted by each factor is of more interest than the total, and we will explore this point further when we discuss the issue of choosing the number of variables. The problem of deciding how much of the total variance should be extracted remains open and subject to the context of a particular analysis.

When a true factor analysis is conducted, the primary objective is to account for the *relationships* among the variables, that is, the shared variability. This objective suggests a different approach to assessing the fit of a solution that involves seeing how well the original correlations can be recovered from the factors. The factors are derived initially from the simple correlations, so the basic idea is to run the analysis backward and see if it is possible to reproduce the original correlations. The closer the reproduced correlations match the original correlations, the better the fit of the solution. Put another way, the smaller the gap between the reproduced and the original correlations—the residuals—the better the fit. How well does our two-factor oblique solution using PAFA meet this criterion?

The original correlation between mental health and general health perception, for example, is .44, and the reproduced correlation is .45: a gap or residual of –.01. The biggest residual occurs for the correlation between energy/vitality and physical function and has a value of –.08. Thus all of the residuals are less than .10, and it turns out that only 5 (17%) are greater than .05. This ability to reproduce the original correlations so closely indicates an excellent fit, given the objective of accounting for the relationships among the variables. Again, note that there is no fixed criterion against which the residuals can be judged. But the potential of this technique for identifying particular variable pairs where the solution does not fit well, and for comparing different solutions in a way that suits the objectives of factor analysis so well, makes it a very useful tool.

The strategy of assessing fit by running the analysis backward can also be applied to a more complex index of correlation called the **communality.** The communality of a variable is the variance it shares with *all* the other variables in an analysis. So it is a multivariate measure of association between one variable and all the others. Any factor analysis begins with a communality value for each variable to represent its shared variance component. In the case of PCA, the communalities are all set to a value of one. This may sound odd, but it is another way of declaring that all variance is available for extraction, that is, all of a variable's variance is treated *as if* it is shared with the other variables. In factor analysis the initial communality for each variable has to be estimated, and this can be achieved in a variety of ways. Usually, it is estimated simply by running a multiple regression in which the variable of interest is treated as a dependent variable and all the others as independent variables. This prior analysis produces an R^2 value for each variable that becomes its initial communality. So, for example, the initial communality for the mental health variable is .68; this variable shares 68% of its variance with the other seven variables.

Just as with the simple correlations, communalities may be reproduced by working backward from the factors in a given solution. Applying this to our two-factor oblique solution using PAFA extraction results in a reassuring picture. For example, the reproduced communality for the mental health variable is .75, giving a residual of −.07. The biggest residual of −.11 occurs for the pain variable, and the other five residuals range between −.02 and −.07. So whether we examine our factor solution from the perspective of simple correlations or multivariate communalities, the criterion of satisfactory reproduction of original associations seems well satisfied.

7.2.4 Choosing the Number of Factors

The number of factors we have extracted was suggested by the proposal that the eight health variables might be classifiable into two categories of mental and physical health. Note that this proposal was not formulated as a strict hypothesis for testing but as a template to be explored. The testing of formal hypotheses about factor structures would take us into the realm of confirmatory factor analysis, which we will investigate later. At the other end of the exploratory-confirmatory continuum, factor analysis is often used to decide how many factors might account for a set of data without any preconceptions on what that number might be. How can the analyst decide on the "right" number of factors?

As usual, there are a number of criteria that need to be balanced out against each other, some of which we have already considered. First and foremost, all of the factors should be *interpretable,* as it is difficult to see what theoretical or practical function could be served by an uninterpretable factor. Jointly, the factors should account for a satisfactory amount of variance (PCA) or shared variance (FA) in the data. As we saw in the last section, this is a guideline in which the notion of "satisfactory" ultimately has to be defined by the analyst. The extracted variance statistics can be used in other ways besides examining the total. Since in the initial extraction each successive factor accounts for a defined proportion of variance, the analyst can decide to keep extracting as long as a factor accounts for at least, say, another 10% of variance. This type of approach can also be expressed in terms of eigenvalues, the extracted variance statistics from which the percentages are derived. A common stopping rule is that a factor should be extracted only if its eigenvalue is at least one. This is equivalent to requiring that any factor should account for at least as much variance as would a single variable. Eigenvalues can also be explored graphically using the **scree plot,** where the analyst looks for a break between the early factors that have large eigenvalues and the later "rubble" of smaller values.

Extracted factors can also be tested for their statistical significance. The particular test depends on the type of extraction procedure: for example, Bartlett's test can be used for PCA or PAFA extraction, whereas other tests, such as Lawley's test are appropriate if maximum likelihood extraction has been conducted. All of the foregoing strategies can be strengthened by cross-validation, that is, checking to see that the number of factors is appropriate across two or more samples. Of course, cross-validation has much wider application than just checking the number of factors and serves as a general assessment of the accuracy of a particular solution. Finally, there is general agreement that overfactoring is preferable to underfactoring. So analysts are often encouraged to explore at least one more factor than these criteria might dictate, and then to reject it if it cannot be justified.

What happens if we apply some of these criteria to our PCA with oblique rotation and extract three factors instead of two? Extracting a third factor would increase the variance accounted for to nearly 76%, which sounds attractive. However, this is only an increase of 8.8%, and the third factor has an eigenvalue of .70, clearly less than one. Most seriously, the pattern matrix of loadings now becomes hard to interpret. The first mental health factor stays intact, but the physical function variable is removed from the second factor to become the only variable loading on the third factor. So, not only is the second factor fractured, but the third factor is ill-defined by only one variable—a so-called singlet. Notice that trying to spread eight variables across three factors is likely to be doomed from the outset since even a roughly uniform spread would produce a factor defined by only two variables. Quite aside from our research objective, there seem to be no grounds for adopting a three-factor solution, and good reasons to accept the two-factor solution. The issues surrounding the appropriate number of factors in a factor analysis are complex and are reviewed in detail by Fabrigar, MacCallum, Wegener, and Strahan (1999, pp. 277–281).

7.2.5 An Example From the Well-Being Literature

After this immersion in some of the details of factor analysis, it is timely to step back and look at an example from the well-being literature. In this area a distinction is made between subjective well-being and psychological well-being. The former focuses on satisfaction with life and the balance between positive and negative emotions. In contrast, the latter focuses on how individuals deal with existential challenges throughout their lives. Keyes, Shmotkin, and Ryff (2002) have suggested that these two conceptions of well-being are "related but distinct" (p. 1009). This implies that a factor analysis of the facets of both constructs should produce two correlated factors, rather than two uncorrelated factors, or just one overall factor. To examine this proposal, Keyes et al. factor analyzed data from a national sample in the United States of about 1,500 adults. The respondents completed measures of three facets of subjective well-being (life satisfaction, positive affect, and negative affect), and six facets of psychological well-being (self-acceptance, environmental mastery, positive relations with others, personal growth, purpose in life, and autonomy).

Keyes et al. report (p. 1011): "Using oblique rotation of principal components extraction, we found that both the scree plot and the eigenvalue-greater-than-one criteria indicated a two-factor structure." This means that extracting any further factors was thought inappropriate because of the small amount of additional variance that would be accounted for. So, even though the factors jointly accounted for only 56% of the total variance, the incremental variance criterion and the theoretical focus of the analysis made this acceptable. After oblique rotation they found that the two factors were correlated .45. All of this was consistent with the "related but distinct" two-factor proposal. To exhibit the relationships among the factors in more detail, Keyes et al. provide the structure matrix, that is, the simple correlations between each of the nine variables and the two factors. Note that since these correlations are not adjusted for the correlation of .45 between the two factors, the overlap in loadings across factors can be seen more clearly than in the adjusted pattern matrix. The three subjective well-being variables loaded only on the first factor, while two of the six psychological well-being variables (personal growth and purpose in life) loaded only on the second. Three of the psychological well-being variables—self-acceptance, environmental mastery, and positive relations—loaded on both factors, while the autonomy variable loaded on neither. Thus the analysis supported the notion of two overlapping factors and provided some insight into the magnitude and nature of the overlap. Having conducted this exploratory analysis, Keyes et al. proceeded to a more formal set of confirmatory factor analyses, which we will discuss in Section 7.4.

7.3 ISSUES OF TRUSTWORTHINESS IN FACTOR ANALYSIS

The trustworthiness of factor analysis results depends primarily on three interrelated issues: the sample size, the nature and quality of measurement, and the relational structure connecting the variables before and after the analysis. Statistical assumptions are of limited concern unless tests of significance are used, which is the exception rather than the rule in exploratory factor analysis.

Fabrigar et al. (1999, p. 274) note that most guidance on sample size in factor analysis takes the form of a recommended minimum number of cases per variable, sometimes accompanied by an absolute minimum. This is doubly problematic because not only do the recommendations vary considerably, but they focus on the number of variables to the exclusion of other relevant considerations. The recommended minimum ratio of cases to variables can vary from 5 to 20, while the absolute minimum can range from 100 to 300. The missing relevant considerations concern the size of the loadings in extracted factors, the number of variables defining a factor, and the size of communalities, that is, the extent to which each variable is tied in to the multivariate structure. Generally speaking, as these three magnitudes increase, fewer cases are needed within the minimal ranges above. Although as Fabrigar et al. comment

(p. 274): "... when these conditions are poor it is possible that samples as large as 400 to 800 might not be sufficient." Stevens (2002, p. 395) provides some useful guidelines on sample size that take most of these considerations into account. It is also worth noting that if statistical tests are to be used in evaluating factors and/or loadings, sample size estimations can be calculated using the usual tools of power analysis.

Factor analysis requires that all variables be measured on at least an interval scale, thereby allowing calculation of the Pearson's *r* correlations that are the starting point of the calculations. As usual, reliable and valid measurement of all variables will help to enhance shared variance (remember that unreliability attenuates correlations) and the interpretation of factors. Clearly, if the variables do not measure what they are intended to, then interpretation of factors may be compromised. Although it is not an issue of measurement quality as such, it is worth reiterating an earlier point here, that factor analyses are more likely to produce acceptable solutions if 3 to 5 variables per expected factor are measured.

The trustworthiness of factor analysis results is dependent on an initial, discernible pattern of relations among the variables that can potentially be resolved into a factor structure. If the variables have little relational structure to begin with, time and energy may be wasted trying to interpret puzzling solutions that are just vain attempts to detect structure where there is none. As we saw at the beginning of Section 7.2, there are various ways to explore the bivariate and multivariate structure of the variables on which the decision whether or not to proceed with a factor analysis can be based.

Unless statistical tests are envisaged, it is not formally required that the data be multivariate normal, linear, and homoscedastic. However, the more these conditions are met, the more trustworthy the correlations will be as a starting point for the analysis. As usual, it is required that the cases are independent of each other. The issue of multicollinearity takes on an interesting aspect in factor analysis because, as Hair et al. (1998, p. 99) note, since highly correlated clusters of variables are sought, some degree of multicollinearity becomes desirable.

To complete this section, it is worth reemphasizing that a trustworthy factor analysis should not necessarily be seen as a definitive one. Even where a factor structure has been validated on another sample, the fact remains that any set of results is the outcome of a series of decisions made by the analyst. The discussion in this chapter should at least make it clear that any factor analysis involves a variety of decisions, any one of which may be treated differently by a different analyst. Only where a clear factor structure appears consistently across a range of decisions and across samples can it be seen as truly trustworthy.

7.4 CONFIRMATORY FACTOR ANALYSIS

Like exploratory factor analysis (EFA), confirmatory factor analysis (CFA) is used to examine the relationships between a set of measured variables and a smaller set of factors that might account for them. As the confirmatory label suggests, though, CFA allows the analyst to specify in advance what these relationships might look like and to test the accuracy of these hypotheses. Often EFA and CFA are used one after the other to generate and then test hypotheses about factor structures using different segments of the same sample of cases. Keyes et al. (2002) used this strategy in the study of subjective and psychological well-being that we introduced in Subsection 7.2.5. Having uncovered a correlated two-factor model using EFA on half of their sample, they then used CFA on the other half of the sample to formally compare the model with a variety of alternatives. We will use these analyses to illustrate the general logic of CFA. As noted earlier, CFA is one application of a family of sophisticated techniques called structural equation modeling. The complexity of these techniques means that our discussion of CFA will be even more cursory than usual and just sufficient to orientate and to indicate its potential. The readings at the end of the chapter will provide a guide to sources for the would-be practitioner.

Any test of a hypothesis has to begin with a precise specification of that hypothesis. In a CFA the first step is to specify a model that comprises a number of hypotheses about the variables to be analyzed. As would be expected, the central hypotheses are about the number of underlying factors, whether or not factors are correlated, and which variables are expected to load on which factors. So, for example, in one of their six well-being models Keyes et al. specified two *uncorrelated* factors with the three subjective well-being variables loading only on one factor, and the six psychological well-being variables loading only on the other factor. Another competing model had the same structure except that the two factors were hypothesized to be *correlated.* If there are more than two factors, CFA, unlike EFA, allows the analyst to specify some factors as oblique (correlated) and others as orthogonal (uncorrelated). The specification of hypotheses can also be thought of as fixing some of the values of model parameters in advance, while leaving others free to be estimated in the analysis. For example, in the first uncorrelated two-factor model, the parameter for the factor correlation was fixed at zero, as were the loading parameters for variables that were *not* expected to appear in a particular factor.

One of the notable strengths of CFA is that, unlike EFA, it enables the analyst to deal explicitly with random measurement error in the variables. At the specification stage, the analyst can hypothesize whether the random errors associated with each variable are correlated or not. Usually, such errors are assumed to be independent of each other, but correlations can occur. For example, if some or all of the variables are measured with a common procedure, such as a self-report 5-point rating scale, errors on one scale may be correlated with errors on another. Measurement error, whether correlated or not, is analyzed in its own right in CFA, and estimates of its presence are a standard part of the results.

Once the model is fully specified, the CFA procedures produce estimates for all of the free parameters in the model, the loadings being those of most interest. The magnitude of the loadings and their statistical significance can then be evaluated. Since the model is intended to reproduce the original relationships among the variables, as in EFA, the residuals or "gaps" between the observed and reproduced correlations can be examined to gauge the overall fit of the model. A variety of goodness of fit measures have been developed using residuals and other approaches, and it is common to find several such measures reported for CFA analyses. The output of a CFA thus allows the analyst to evaluate a factor model overall and at the level of individual variable-factor relationships. Typically, the evaluation is comparative, pitting models against each other.

Of the six models evaluated by Keyes et al., two became the focus of particular interest. Their Model 4 specified two oblique factors with all variables loading on one factor or the other. Model 6 also specified two oblique factors, but with two variables—self-acceptance and environmental mastery—loading on both factors. Thus the two models provided two different interpretations of the proposed "related but distinct" nature of subjective and psychological well-being. In terms of overall fit, Model 6 was shown to be superior to Model 4 on five different fit measures. The two factors were correlated .84 in Model 4 and .70 in Model 6. However, the pattern of loadings in the two models led the authors to favor Model 4, although they noted that both had heuristic value. In particular, they noted that the loadings for the self-acceptance and environmental mastery variables were strikingly lower in Model 6 than in Model 4, partially undermining the structure of both factors. They concluded: "In sum, although Model 6 demonstrates the partial overlapping and, hence, the high common variance between subjective and psychological well-being conceptions, Model 4 more clearly presents them as correlated yet distinct in content" (Keyes et al., 2002, p. 1013). This study provides a nice illustration of the interplay among theory, comparative model fit, and loadings that CFA allows.

The Keyes et al. study shows how CFA can be used to investigate a theoretical structure. It can also be used to investigate psychometric issues, as illustrated by research on a measure of quality of life: the World Health Organization Quality of Life Assessment (WHOQOL). This complex measure was developed and tested collaboratively in 15 countries, and the first major report of its psychometric properties appeared in 1998 (WHOQOL Group, 1998). The 100- item WHOQOL measure assesses 24 "facets" of quality of life such as positive feelings, self-esteem, social support, and mobility, as well as an overall perception. One of the questions of interest to the developers was whether these facets could be organized into "domains" and, if so, how many were appropriate. A conceptual analysis suggested the presence of six domains, while exploratory principal components analysis with varimax rotation suggested four. Accordingly, the developers used CFA to compare models with four and six domains, respectively, and included a third model that specified just one all-encompassing domain. They found that the four-domain model produced the best fit to the data in separate samples of ill and well participants. Moreover, they noted that fit was improved further when the measurement error for some of the facets was specified as correlated. Overall, this study shows how CFA can uncover simple structure in highly complex data using systematic hypothesis testing.

The use of CFA raises many technical issues for the analyst, such as deciding whether correlations or covariances should be the starting point of the analysis, when variables should be standardized, what the appropriate number of free versus fixed parameters are (the identification problem), and which estimation technique should be used. It should also be emphasized that CFA can be applied to other sorts of problems than those illustrated here, such as comparing factor structures across samples: multiple group models or simultaneous CFA. Guidance on these and other issues is available in the readings cited in the next section.

7.5 FURTHER READING

The text by Gorsuch (1983) is still regarded by many as the best source of guidance to factor analysis. Briefer accounts can be found in Kim and Mueller (1978a, 1978b) and, more recently, Fabrigar et al. (1999). Stevens (2002, Chapter 11) provides a good overview of both exploratory and confirmatory factor analysis, as do Tabachnick and Fidell (2001, Chapters 13 and 14). Confirmatory factor analysis is one form of structural equation modeling that can be explored further in Schumacker and Lomax (1996).

EIGHT

LOG-LINEAR ANALYSIS

In this final chapter we explore log-linear analysis (sometimes called multiway frequency analysis): a technique for analyzing categorical data when all variables are treated as if they were independent variables. Log-linear analysis in its most general form is used to detect the pattern of relationships in a network of categorical variables. Since the variables are categorical, this type of analysis is inherently one of *group* differences. In log-linear terms, a relationship between a pair of variables is called a two-way interaction, that among three variables, a three-way interaction, and so on. The general aim is to find which associations are present and, in particular, to arrive at the most parsimonious picture. In a network of four variables, for example, there may be up to 1 four-way, 4 three-way, and 6 two-way interactions. The ability to exclude with confidence any of these eleven possible relationships from further consideration is a distinct gain in the constant scientific drive to derive simple pictures from complex data. This process may be exploratory, in which case the data determine the outcome; or it may be confirmatory, whereby analysts explicitly test their hypothesized models of associations against the data.

From an accounting perspective, the general goal of log-linear analysis is to account for the multivariate distribution of a set of categorical variables. This multivariate distribution can be displayed in a contingency table. As we saw in Chapter 1, when there are two variables, the contingency table has two dimensions—rows and columns—and the joint distribution of the two variables is shown by the number of cases in each cell of the table. Each time a variable is added, another dimension is required in the table, and the structure quickly becomes hard to visualize. But the multivariate distribution continues to be evident in the frequencies found in the cells that result from cross-classifying the variables. For example, if there are four dichotomous variables, cross-classification will produce $2 \times 2 \times 2 \times 2 = 16$ cells, each with its own subset of cases. In log-linear analysis these frequencies are treated as the values of a sort of dependent variable, which need to be accounted for. So the broad analytic question is: Which combinations of (independent) variables are responsible for the configuration of cell frequencies?

One of the most powerful strategies within the log-linear family is called **hierarchical log-linear analysis,** the technique that will occupy us for most of this chapter. This approach envisages a hierarchy of associations, whereby simpler relations are nested inside more complex ones. So, for example, in the four-variable network, the three-way associations are seen as "contained" within the four-way association, and the two-way associations are "contained" within the set of three-way associations. This hierarchical structure creates a framework within which statistical tests of associations can be carried out in a highly systematic fashion. In Section 8.1 we examine hierarchical log-linear analysis, first using the nurses' data and then with an example from the research literature. As usual, we will focus on the nature and derivation of the composite variable in this context and on the statistics that are used to address the questions of interest in this type of analysis. In Section 8.2 we discuss the few issues surrounding the trustworthiness of results from a log-linear analysis. Then, in Section 8.3 we explore another type of log-linear analysis called **logit analysis.** In this type of categorical analysis, we return to the situation where one of the variables is treated as dependent, and the aim is to account for differences on this variable using the other variables in the set. Since, more specifically, the aim is to predict the odds on the dependent variable, it will come as no surprise to learn that logit analysis is the special case of logistic regression (see Chapter 5) in which all variables are categorical. Once again, we will find the apparent multiplicity of multivariate techniques simplified as one technique collapses into another.

8.1 HIERARCHICAL LOG-LINEAR ANALYSIS

Log-linear analysis can seem rather daunting because of its level of abstraction and because of its terminology, which sometimes gives a different meaning to terms that are familiar from other statistical contexts. Accordingly, it will be

Table 8.1 Contingency Table Showing Relationships Among Control, Autonomy, Relations, and Health

		Poor Relations		*Good Relations*		
		Poor Health	*Good Health*	*Poor Health*	*Good Health*	
Poor Control	Poor Autonomy	28	19	5	7	59
	Good Autonomy	4	11	3	11	29
Good Control	Poor Autonomy	2	6	3	5	16
	Good Autonomy	11	6	16	26	59
Total		45	42	27	49	163

helpful to ground the discussion of hierarchical log-linear analysis in a familiar example. Returning to the nurses' data, we will use log-linear analysis to explore the associations among four variables: professional relations, autonomy, control, and mental health. Remember that for this type of analysis all variables have the same status, rather than being divided into independent and dependent sets. The variables are best seen as independent in that they are used to account for the frequency distribution in the contingency table. Since log-linear analysis requires categorical data, all four variables have been dichotomized, with the two categories for each variable labeled "poor" and "good." As we have noted on earlier occasions, collapsing scores into categories is not recommended unless the distribution of the scores suggests that categories are preferable. The creation of categories here is motivated simply by the desire to continue using a familiar data set that happens to have few intrinsically categorical variables.

If we cross-classify the four dichotomous variables, we create a contingency table with 16 cells, which appears in Table 8.1. Since the variables all have equal status, the way in which the categories are configured is arbitrary.

The numbers in the 16 cells are the observed conditional frequencies that show how the 163 cases are distributed according to (that is, conditional on) their joint categories. Thus the 28 cases in the top-left cell are categorized as poor on all four variables; 7 cases in the top-right cell are categorized as in good mental health, while reporting good relations but poor control and poor autonomy; and so forth. These observed conditional frequencies have been summed across rows and columns to show the observed marginal frequencies. Strictly speaking, the term "marginal frequencies" refers to those that indicate the distribution on each variable separately, and these are not shown. They are easily calculated by summing across appropriate categories. For example, the relations distribution appears in the bottom row: 45 + 42 = 87 cases had poor relations, while 27 + 49 = 76 had good relations. The autonomy distribution is embedded in the right-hand column: 59 + 16 = 75 cases had poor autonomy, and 29 + 59 = 88 cases had good autonomy. These univariate marginal frequencies will have a role to play in the analysis, but they have been omitted from the table to keep the focus firmly on relationships among the variables and to enhance clarity.

Before the development of log-linear analysis, the associations among a set of categorical variables were analyzed by collapsing or slicing the contingency table. These strategies made it possible to then test pairwise associations with the chi^2 test of independence that we discussed in the first two chapters. For example, the (two-way) relationship between autonomy and control could be analyzed using a four-cell table created by collapsing the categories of relations and mental health. Or the three-way association among autonomy, control, and relations could be analyzed by testing the two-way association between autonomy and control for each category of relations. Earlier discussions on the perils of viewing multivariate data through bivariate spectacles should make another lengthy explanation redundant. In summary, the multiple bivariate tests offer no control of confounding and run the risk of inflating Type I error. Further, the analysis of interaction effects involving three or more variables should provide tests of these effects as such, not just a series of bivariate tests within categories of other variables. Once again, to obtain this sort of analytic power, it is necessary to turn to multivariate analysis, in this case log-linear analysis.

8.1.1 The Composite Variable in Log-Linear Analysis

The cell frequencies in Table 8.1 have been generated by the four variables, acting singly and in all possible combinations. The complete set of variables and their associations is known as a **saturated model** since no possible effect is excluded. If we were asked to say what cell frequencies we would expect to find in the table if the saturated model were correct, the answer would simply be those that actually do appear in the table. In other words, if all of the variables have all possible effects, the observed and expected frequencies are identical. What would we expect the frequencies to look like at the opposite extreme, where none of the variables had any effects? In this situation the cell frequencies would all be the same, that is, each would be $163/16 = 10.19$: a constant value. (These expected frequencies are hypothetical, so we need not worry about requiring that cells contain whole cases.) The saturated and constant models define two uninformative extremes: Either all of the variables are related in every conceivable way or they are totally unrelated. However, we can make good use of this strategy of generating expected frequencies if we choose models between these two extremes, where only *some* of the possible effects are present. Broadly speaking, we will build models with selected effects, generate the frequencies that would be expected if the model were correct, and then compare the expected with the observed or actual frequencies to see how accurate our expectations are.

In log-linear analysis, an expected frequency is calculated for each cell of the contingency table. If there are associations among the variables, they will create inequalities in the expected cell frequencies, that is, deviations from the constant value expected when the variables have no effects. These expected frequencies are calculated so that they reflect the particular effects included in the model of interest. Then the question is how well the expected frequencies map onto the observed frequencies. All of this is achieved using the maximum likelihood estimation technique that we first encountered in Chapter 5. The set of expected and observed frequencies can be compared using the likelihood ratio chi^2, which grows smaller as the gap between the expected and observed frequencies lessens. So, broadly speaking, the strategy is to find the expected frequencies that minimize the likelihood ratio chi^2, or in other words, that maximize the likelihood of reproducing the observed frequencies. As in the case of logistic regression, this process is iterative, cycling repeatedly through the various constraints imposed by the selected model until the estimates of the expected frequencies stabilize.

With this overview, we can now turn to the question of what the composite variable looks like in log-linear analysis. As the name of the technique suggests, a linear composite can be achieved if we work in log frequencies rather than in frequencies. The composite variable in this context is therefore:

log expected frequency = [coefficient 1(effect 1)
+ coefficient 2(effect 2) + ...
+ constant]

Instead of a predicted score on the dependent variable we have an expected frequency for a cell. As in logistic regression, if we work with the natural logs of frequencies, the composite variable takes on the familiar form of a weighted sum. The coefficients are known here as **parameter estimates,** which can be calculated from the expected frequency estimates derived earlier. There is a coefficient for each effect included in the model of interest. So for the saturated model of our four dichotomous variables, as we saw earlier, there would be eleven effects representing two-, three-, and four-way relationships, plus the four individual effects of the variables, giving a total of 15 effects and coefficients. Note that although these are referred to as coefficients, there are no independent variable scores by which they are multiplied since there are no scores. However, they bring the usual benefits of multivariate coefficients. They are partial coefficients, showing the relative magnitude of an effect while controlling for other effects in the model. Further, the null hypothesis that a particular coefficient is zero in the population sampled can be evaluated with a *z* test statistic. There will be a composite for every cell of the contingency table—16 in our example—although constraints built into the estimation of expected frequencies and coefficients mean that the values of many coefficients are mutually determined. The observed and expected frequencies, the likelihood ratio chi^2 statistic, and the parameter estimates and associated *z* tests provide a useful set of tools, as we can now illustrate.

8.1.2 Log-Linear Analysis in Action

We will now conduct a log-linear analysis of the data in Table 8.1. The primary objective is to discover in what ways the four variables are related to each other. Note that this objective is exploratory; we have no predictions about what associations may be present. In this circumstance the saturated model containing all possible effects is a good place to start exploring. When this model, with all 15 possible effects, is analyzed, the following coefficients and *z* statistics result. In Table 8.2 each variable is represented by its initial letter so that the effects are easily labeled.

If we assume an alpha of .05, a coefficient can be considered statistically significant when the associated *z* value exceeds 1.96, though some recommend a threshold *z* of 4.00 to compensate for the multiple tests (Tabachnick & Fidell, 2001, p. 241). By the more liberal criterion, the table shows that there are 1 one-way, 3 two-way, and 1 three-way effects that are unlikely to be due to

Table 8.2 Hierarchical Log-Linear Analysis of the Relationships Among Control, Autonomy, Relations, and Health

Effects		*Coefficients*	*z Values*
Four-way			
	C X A X R X H	.124	1.27
Three-way			
	C X A X R	.098	.48
	C X A X H	.231	2.37
	C X R X H	.025	.26
	A X R X H	–.069	–.71
Two-way			
	C X A	.441	4.51
	C X R	.299	3.06
	A X R	.247	2.52
	C X H	–.047	–.48
	A X H	.043	.44
	R X H	.093	.95
One-way			
	C	.096	.99
	A	–.141	–1.44
	R	.057	.59
	H	–.211	–2.16

NOTE: C = control; A = autonomy; R = relations; H = health

chance. How can these be interpreted? The one-way effect for health simply indicates that the distribution of cases across the poor and good categories is sufficiently unequal to produce an odds significantly different from 1. According to Table 8.1, there are 45 + 27 = 72 cases in the poor category, and 42 + 49 = 91 cases in the good category, and so an odds of 72/91 = .79 of being in poor health. The interpretation is clear, but of little interest given our concern with associations, that is, at least two-way effects.

A two-way effect simply indicates that two variables are associated. The results of this analysis show that each pair of the workplace variables—control, autonomy, and relations—are correlated to a degree that is unlikely to be due to chance. Put another way, the odds ratio for each pair is significantly different from 1. Note also that each of these two-way effects is calculated while holding the others constant, so we can have some confidence that the effects are not confounded with each other. Moreover, the coefficients are comparable, so we can conclude that the association between control and autonomy is clearly the strongest of the three, while the other two are of comparable magnitude.

This is also evident if we calculate the odds ratio for the three associations using the appropriate frequencies from Table 8.1. The odds ratio for control and autonomy is (59/29)/(16/59) = 7.5, indicating that nurses who are in the good autonomy group are 7.5 times as likely to be in the good control group than those in the poor autonomy group. The odds ratio for control and relations is (50/26)/(25/62) = 4.8, and that for autonomy and relations is (56/20)/(32/55) = 4.8.

Table 8.2 also shows that there is a statistically significant three-way effect for control, autonomy, and health. It is important to appreciate that the order of variables is arbitrary here since all of the variable have the same statistical status. Whichever order we place them in, we are still referring to the same three-way interaction effect. For interpretation purposes, we can superimpose any conceptual framework that is appropriate to the analytic context. For example, we can temporarily treat health as a dependent variable, autonomy as an independent variable, and control as a moderating variable. Within this framework, the three-way effect can be described as showing that the effect of autonomy on health differs according to the level of control. This means that the odds ratio for autonomy and health in the poor control group is significantly different from that in the good control group. In the poor control group the odds ratio is (22/26)/(7/33) = 3.99, while in the good control group the odds ratio is (32/11)/(27/5) = .54. When these two associations are tested with the likelihood ratio chi^2, the first is statistically significant ($p = .004$), but the second is not ($p = .292$). These follow-up analyses suggest that good autonomy is associated with good health, but only if control is poor. If control is good, the level of autonomy has no health implications. It is important to reiterate that this is only one way to view the three-way effect and would need to be defended on nonstatistical grounds.

As we noted earlier, the use of hierarchical models enables the analyst to systematically test effects in a variety of ways. Most of these involve the comparison of subsets of effects or submodels and use the *change* in the likelihood ratio chi^2 between two models as the basis for tests. One particularly useful testing procedure is known as the **test of partial association.** Essentially, this tests for the statistical significance of an effect by comparing two models, one with the effect present and one with it absent. The difference between the likelihood ratio chi^{2}s for the two models provides a partial chi^2. If that chi^2 is significant, it is treated as evidence that the effect makes a contribution to the model; excluding the effect makes a significant difference. The tests of partial association for the nurses' data produce four partial chi²s that are statistically significant. Reassuringly, they are the three-way effect for control, autonomy, and health ($p < .008$), and the two-way effects for control and autonomy ($p < .0001$), control and relations ($p < .002$), and autonomy and relations ($p < .005$). If an effect is apparent from a variety of testing perspectives, we can have increased confidence in its presence. In the present example the evidence from the tests of the coefficients and from the tests of partial association converges in a satisfying way.

We have identified 1 three-way and 3 two-way effects, and it is tempting to conclude that these are all we need to include in a final model that accounts for the data as well as possible. However, we have to remember that all of the testing has been conducted within a hierarchical framework that cannot be jettisoned at the end of the process. The practical implication of this is that, when an effect is retained, so are any effects that are nested inside it. So, choosing to include the three-way effect in our final model entails the inclusion of all the two- and one-way effects that can be constructed from these three variables. Similarly, the inclusion of any two-way effect entails the inclusion of its two component one-way effects. All of this can quickly become confusing, so it is helpful to have a shorthand system based on the notion of a **generating class.** The generating class of a model is the set of actual and implied effects contained in the model. If we use the initial letters of the four variables from Table 8.2, the generating class of the saturated model would be C X A X R X H. This says that the model comprises a four-way effect and all of the three-, two-, and one-way effects that can be formed from the four variables. The generating class for the unsaturated model that we have arrived at is C X A X H, C X R, and A X R. Note that the C X A effect does not need to be specified separately because it is already nested inside the three-way effect. Similarly, all of the one-way effects are included by implication.

When log-linear analysis is used in exploratory mode, it is also possible to trawl for associations, using automated statistical criteria. For example, in the SPSS program the search for associations among a set of categorical variables can be conducted using a backward elimination strategy. This is the same type of backward model building that we discussed in Chapter 4 in the context of multiple regression. The program starts with the full or saturated model and then uses partial chi² tests to discover whether the elimination of an effect makes a significant difference. The effects are tested in order of decreasing complexity, and an effect is removed if its partial chi² is *non*significant, that is, its disappearance results in no appreciable loss. Not surprisingly, given this strategy, a backward elimination analysis of the nurses' data resulted in the same model that we uncovered by inspection of the saturated model. Note that this unsaturated model with generating class C X A X H, C X R, and A X R contains both significant and nonsignificant effects because of its hierarchical structure. The presence of a few nonsignificant effects is, however, a small price to pay for the analytic power of hierarchical modeling. Note also that the use of automated, statistically driven log-linear analysis is subject to all of the problems we reviewed in Chapter 4.

Although, as we have seen, the saturated model is a useful exploratory tool for identifying effects, it does not offer the potential for *overall* model evaluation. Since the expected frequencies from a saturated model simply reproduce the observed frequencies, the question of how well the model accounts for the data cannot be addressed. There will be no residual gaps between the observed and expected frequencies, and so the likelihood ratio chi^2 value that summarizes them will be an uninformative zero. To see how a log-linear model can be evaluated, we will now analyze the unsaturated model with the generating class C X A X H, C X R, and A X R.

The likelihood ratio chi^2 value for this model as a whole is 5.38 with an associated p value of .372. This looks like bad news, but on this occasion the opposite is true. Now we are using a chi^2 statistic to evaluate how well a model fits the data. The better the fit, the smaller the residuals, and the smaller the chi^2 value. So, for once, we are hoping for a small test statistic value that is *not* statistically significant. This backwards approach to p values creates a problem in deciding what value an analyst might use to decide that a model is acceptable. The solution to the problem ideally involves careful consideration of Type I and II errors and of sample size for a particular analysis. But, as a general guide, Knoke and Burke (1980, p. 31) suggest that an acceptable range for the p value would generally lie between .10 and .35. So, on this criterion our model fit is marginally acceptable with a slight risk of being overinclusive in its selection of effects.

In practice, models tend to be pitted against each other, so attention focuses as much on how the p value of the likelihood ratio chi^2 changes from model to model, as on its absolute value. For example, in the penultimate step of the backward model selection described above, the model included all of the effects in the final model plus the two-way effect for relations and health. The chi^2 statistic for this model has a p value of .725: a clear signal that at least one unnecessary effect has been retained. Conversely, if we remove the significant three-way effect from our chosen model, the p value for the model's chi^2 sinks to .055. These three p values of .372, .725, and .055 suggest that our chosen model is the best choice of the three tested. That said, it should be stressed that all of our discussion has been framed in terms of exploratory analysis. When log-linear analysis is used to test effects that are hypothesized in advance, the overall model fit is of less concern. Further, as we have seen on a number of occasions, p values are only one type of criterion that analysts use to make decisions, and one that has its costs and benefits.

The question of how well a model accounts for a set of data can be addressed in more detail by examining the residuals themselves, cell by cell. The residuals are the differences between the observed frequencies and the expected frequencies generated by the model. The size and signs of residuals, and the cells where they occur, can provide useful clues on which subgroups of cases are poorly accounted for by the model. In the analysis of our chosen model, the largest residual is 3.5 and it occurred in the cell for cases with good control, good autonomy, poor relations, and poor health. In the sample, 11 cases fell into this cell, but the model predicted only 7.5. This may not sound too impressive, but we can gain a broader perspective by standardizing the residuals. Standardized residuals are z statistics, which means that they can be compared with a value of ± 1.96 as a rough guide to deciding whether they are large enough to warrant concern (Tabachnick & Fidell, 2001, p. 264). In the present case the residual of 3.5 translates into a z value of 1.30, so there seems to be no basis for suggesting that the model is a poor fit either overall, or cell by cell.

To complete this section, we will return to the study on religious attendance and life satisfaction conducted by Hintikka (2001), which we discussed in Chapter 5. Using logistic regression on data from a random sample of Finnish adults, Hintikka showed that religious attenders were 1.7 times more likely to be satisfied with their lives than nonattenders. This effect was present when employment status, household status, and social support were statistically controlled. Hintikka also conducted a hierarchical log-linear analysis of these five variables. All variables were categorical, but unlike in the logistic regression, religious attendance and life satisfaction were treated as trichotomies rather than dichotomies. As this suggests, log-linear analysis can cope with any number of categories, not just the dichotomies we have focused on for the sake of clarity.

Hintikka briefly reports the results for a "nearly complete" model that is otherwise unspecified. Since the model was unsaturated and hierarchical, we can deduce that at least the five-way effect was not included. The chi^2 for this model had a p value of .99, indicating an acceptable fit. Since this is described as the "final model," we can also deduce that other models were also explored. The following significant effects are reported: employment status X household status X social support X life satisfaction; religious attendance X household status X life satisfaction; religious attendance X employment status; and religious attendance X social support. In the context of the study objective, the two-way effects are of little interest. They simply showed that religious attendance was associated separately with employment status and with level of social support. However, the three-way and four-way effects are more interesting because they extend the findings from the logistic regression, which did not include interaction effects.

The three-way effect for religious attendance, household status, and life satisfaction may help to explain why the log-linear analysis did not replicate the association between religious attendance and life satisfaction from the logistic regression. One interpretation of the three-way effect would be that religious attendance was associated with life satisfaction, but that this association differed according to household status. Moreover, maybe the difference was such that it canceled out the overall association between religious attendance and life satisfaction. Of course, this is speculative pending more detailed analysis. An alternative explanation for the different logistic and log-linear results may just be the use of dichotomies and trichotomies, respectively. For present purposes, the objective is not to reach definitive conclusions but to show log-linear analysis in action. The four-way effect is also interesting even though it does not include religious attendance. It suggests that the level of life satisfaction was partly determined by employment status, household status, and level of social support acting jointly in some complex fashion. Teasing out the way in which this occurred is beyond the scope of this discussion, and it is not pursued in Hintikka's article. But the presence of such an effect may have interesting implications for theory and for what sorts of statistical control need to be exercised in future analyses.

8.2 TRUSTWORTHINESS IN LOG-LINEAR ANALYSIS

Of all of the trustworthiness sections in the book, this can be the shortest since so few requirements need to be met for a log-linear analysis to be legitimate.

Most of the requirements are concerned with aspects of sampling, which are clearly discussed by Tabachnick and Fidell (2001, p. 223). They recommend that the total sample have at least five times the number of cases as there are cells in the analysis. In our example the presence of 16 cells indicates a minimum required sample size of 80, under half of the number actually used. In passing, it is worth noting how quickly the number of cells, and therefore required cases, multiplies with the addition of categories to variables and of variables to the model. Even if we had restricted ourselves to four variables but treated each as a trichotomy rather than a dichotomy, there would have been $3 \times 3 \times 3 \times 3 = 81$ cells, and the required minimum number of cases would have leapt to $81 \times 5 = 405$.

Even if the total sample size is adequate, the very nature of cross-classification means that some cells will have very few cases or perhaps none at all. So the *distribution* of observed frequencies can also be an issue. In our example, despite the total number of cases relative to cells, one cell had only two cases in it, while three others had fewer than five cases. If too many cells are empty, the analysis may fail to converge on stable estimates for the expected frequencies. There are various strategies that can be considered to deal with empty cells, but no easy solutions. A common strategy is to routinely add a constant, such as .5, to each cell before the analysis begins. Although this is common practice, Tabachnick and Fidell recommend against it on the grounds that it reduces statistical power. Instead they recommend collapsing categories or deleting variables where this can be justified on theoretical and statistical grounds. A special situation arises if a cell is empty because it is impossible for it to contain cases under any circumstances. For example, if absence or presence of postnatal depression is cross-classified with gender, there will be no cases in the cell labeled "present/male" (though some fathers may disagree). In this type of situation the cell may be designated as containing what is called a **structural zero,** and the analysis "works around" it.

Another reason for being careful about the number of cells in a log-linear analysis concerns the expected frequencies generated by the model. If there are too many small expected frequencies, the power of the analysis can be seriously compromised. Tabachnick and Fidell recommend that chi^2 analyses be carried out on all possible two-way contingency tables to generate expected frequencies. All expected frequencies should be greater than one, and no more than 20% of the cells should contain expected frequencies with values less than 5. Where these criteria are not met, the solutions are the same as those for combating inadequate observed frequencies. Naturally, the two types of frequencies are related, so successfully attending to the observed frequencies will often avoid problems with the expected frequencies.

In its most basic form, log-linear analysis requires categorical variables where each case appears in only one category of each variable. Accordingly, the grand total of the observed frequencies across all categories should be equal to the sample size. Log-linear analysis can also accommodate ordinal variables, but a discussion of this lies beyond the scope of a brief introduction. Whatever the variables, it is assumed that they are measured with adequate reliability and validity. As we noted in Chapter 5, measuring complex phenomena on a categorical scale may be no less demanding than doing so with more elaborate scales, and the consequences of doing it badly are no less serious.

Assumptions underlying the tests used in log-linear analysis are almost nonexistent. It is assumed that the cases in each cell are independent of each other, as usual. Otherwise, there is little to be said, except for a natural concern to identify and deal with influential outliers in the form of large residuals, and to appreciate the need for cross-validation on other samples to avoid overcapitalization on chance findings. This in turn raises a final issue of trustworthiness that we have discussed in detail elsewhere, namely the heavy reliance on *p* values. It should be clear from even this brief introduction that log-linear analysis makes extensive use of *p* values as a way of identifying trustworthy effects and models. The multiple, often overlapping, tests of competing models, create *p* values that will rarely indicate the true level of Type I error. Accordingly, it is advisable to treat such *p* values with caution and, as ever, to view them as just one of a set of theoretical and statistical criteria by which the trustworthiness of results should be judged.

8.3 LOG-LINEAR ANALYSIS WITH A DEPENDENT VARIABLE: LOGIT ANALYSIS

Log-linear analysis is a family of techniques. The main focus of this chapter has been on general hierarchical log-linear analysis since it addresses the less usual situation where no distinction is made between independent and dependent variables. In this last section we turn to another family member—logit analysis—which reinstates the independent-dependent variable distinction. All of the procedures we have discussed in this chapter can also be used to analyze how a set of categorical independent variables account for a categorical dependent variable. Nothing changes except that now, instead of using the composite variable to generate expected frequencies, we use it to generate the *odds* of the expected cell frequencies for the dependent variable. As before, all of this is done in natural logs to produce an additive model. A logit model contains a constant, a term for the dependent variable, and terms for whatever independent-dependent variable associations are proposed by the analyst. Each of these latter terms will represent either a simple effect of an independent variable or an interaction effect involving two or more independent variables. Although it is not apparent from the terms in a logit model, the analysis also controls for all of the effects *within* the set of independent variables. This parallels the situation in multiple regression where correlations among the independent variables are controlled even though they do not appear as such in the composite variable. So, in sum, logit analysis can be used to test the effects of a set of independent variables on a dependent variable while controlling for all confounding associations in the model.

To make all of this more concrete, we will explore a study conducted by Benin and Nienstedt (1985) on the determinants of happiness in four national samples in the United States: 774 housewives, 472 husbands of housewives, 922 working wives, and 882 husbands of working wives. Using data from the National Opinion Research Center's General Social Surveys for 1978–1983, they analyzed the effects of job satisfaction (three categories), marital happiness (two categories), stage of life (four categories), and education (two categories) on overall happiness for each sample. The happiness classification had three categories (very, pretty, and not so happy), but the researchers collapsed these categories to form dichotomies. We will focus on their analyses of the very happy category versus the other two. It is noteworthy that this configuration of variables produces a contingency table with 96 cells. If we apply Tabachnick and Fidell's criterion of at least five cases per cell, the required sample size for each analysis would be 480, which renders the husbands of housewives sample of 472 barely adequate.

Separate logit analyses were conducted for each of the four samples. The analytic strategy was a forward stepwise procedure whereby terms representing independent-dependent variable associations were entered into the model sequentially and retained if they made a statistically significant contribution. In their article Benin and Nienstedt report for each of the four samples the chi^2 and p values for the resulting model, and the parameter estimates and odds ratios for the statistically significant effects. In ascending order, the p values for the four models were .30, .69, .96, and .99. All but one of these exceeds Knoke and Burke's (1980) recommended range of .10 to .35, which raises the possibility of "overfitting" due to an excess of terms. However, the chosen models were the best fitting of those tested and, as we will see, they actually contained few effects.

For housewives and for husbands of housewives, marital happiness and job satisfaction were independent correlates of overall happiness. For example, in the housewives sample, wives who were very happily married were more than three times as likely to report being very happy overall, compared with those not so happily married. The corresponding odds ratio in the husbands of housewives sample was 2.12. Although marital happiness and job satisfaction influenced overall happiness in the samples of working wives and husbands of working wives as well, the effects were interactive rather than independent. Thus for both groups there was a three-way effect of similar magnitude—happiness × marital happiness × job satisfaction—but no constituent two-way effects. In these groups, enhanced overall happiness was dependent on the *combination* of marital happiness and job satisfaction; marital happiness was influential but only if respondents were also satisfied with their job, and vice versa.

Finally, the analyses also uncovered another intriguing two-way association whereby life stage had an independent effect on overall happiness in the two samples of husbands, but not in either of the wives samples. The researchers were able to use parameter estimates and odds ratios to show further that the pattern of the life stage effect differed between the two samples of husbands. The husbands of housewives appeared to be most happy before any children were born and least happy when the children were preschool or at school. In contrast, the husbands of working wives were least happy when children were school age and most happy when they had left home. All in all, the study is a good example of how log-linear analysis can be used to find interesting patterns in complex data, even when the data are the product of relatively simple measurement processes.

The family of log-linear techniques has many more applications than those presented here. As we noted earlier, variables with an ordinal scale can be included. If a logit model includes interval-level independent variables, another family member, logistic regression, can be used. So the logit analysis we have just reviewed could have been conducted using logistic regression with precisely the same results. Logit analysis is the special case of logistic regression where all variables are categorical. Log-linear techniques can also be used to analyze data sets where there are multiple independent and multiple dependent variables. Causal models with mediating relationships can be analyzed, as can data from longitudinal studies. Such techniques are beyond the scope of this book, but this basic introduction to the fundamental notion of a log-linear model should provide a useful framework within which more sophisticated understandings could be developed.

8.4 FURTHER READING

Knoke and Burke (1980) provide a brief but accessible introduction to log-linear analysis, as does Rodgers (1995). More extensive introductory treatments with computer-based analyses can be found in Tabachnick and Fidell (2001, Chapter 7), and in Stevens (2002, Chapter 14). A text-length account that is less technical than others is available in Wickens (1989).

BIBLIOGRAPHY

Achen, C. H. (1982). *Interpreting and using regression.* Thousand Oaks, CA: Sage.

Baron, R. M., & Kenny, D. A. (1986). The moderator-mediator variable distinction in social psychological research: Conceptual, strategic and statistical considerations. *Journal of Personality and Social Psychology, 51,* 1173–1182.

Benin, M. H., & Nienstedt, B. C. (1985). Happiness in single- and dual-earner families: The effects of marital happiness, job satisfaction, and life cycle. *Journal of Marriage and the Family, 47*(4), 975–984.

Bjorkman, T., & Hansson, L. (2002). Predictors of improvement in quality of life of long-term mentally ill individuals receiving case management. *European Psychiatry, 17,* 33–40.

Budge, C., Carryer, J., & Wood, S. (2003). Health correlates of autonomy, control and professional relationships in the nursing work environment. *Journal of Advanced Nursing, 42*(3), 260–268.

Cohen, J. (1988). *Statistical power analysis for the behavioral sciences* (2nd ed.). Hillsdale, NJ: Lawrence Erlbaum.

Cohen, J. (1990). Things I have learned (so far). *American Psychologist, 45*(12), 1304–1312.

Cohen, J. (1992). A power primer. *Psychological Bulletin, 112*(1), 155–159.

Cohen, J. (1994). The earth is round ($p < .05$). *American Psychologist, 49*(12), 997–1003.

Cohen, P., Cohen, J., West, S. G., & Aiken, L. S. (2003). *Applied multiple regression: Correlation analysis for the behavioral sciences* (3rd ed.). Hillsdale, NJ: Lawrence Erlbaum.

Darlington, R. B. (1990). *Regression and linear models.* New York: McGraw-Hill.

Diehl, M., Elnick, A. B., Bourbeau, L. S., & Labouvie-Vief, G. (1998). Adult attachment styles: Their relations to family context and personality. *Journal of Personality and Social Psychology, 74*(6), 1656–1669.

Emmons, R. A., & McCullough, M. E. (2003). Counting blessings versus burdens: An experimental investigation of gratitude and subjective well-being in daily life. *Journal of Personality and Social Psychology, 84*(2), 377–389.

Fabrigar, L. R., MacCallum, R. C., Wegener, D. T., & Strahan, E. J. (1999). Evaluating the use of exploratory factor analysis in psychological research. *Psychological Methods, 4*(3), 272–299.

Frick, R. W. (1996). The appropriate use of null hypothesis testing. *Psychological Methods, 1,* 379–390.

Gorsuch, R. L. (1983) *Factor analysis* (2nd ed.). Hillsdale, NJ: Lawrence Erlbaum.

Green, S. B. (1991). How many subjects does it take to do a regression analysis? *Multivariate Behavioral Research, 26,* 499–510.

Hair, J. F., Anderson, R. E., Tatham, R. L., & Black W. C. (1998). *Multivariate data analysis* (5th ed.). Englewood Cliffs, NJ: Prentice-Hall.

Hayes, N., & Joseph, S. (2003). Big 5 correlates of three measures of subjective well-being. *Personality and Individual Differences, 34*(4), 723–727.

Hays, W. L. (1994). *Statistics* (5th ed.). Stamford, CT: Thomson.

Hintikka, J. (2001). Religious attendance and life satisfaction in the Finnish general population. *Journal of Psychology & Theology, 29*(2), 158–164.

Huberty, C. J. (1984). Issues in the use and interpretation of discriminant analysis. *Psychological Bulletin, 95*(1), 156–171.

Huberty, C. J. (1994). *Applied discriminant analysis.* New York: Wiley.

Jaccard, J. (1998). *Interaction effects in factorial analysis of variance.* Thousand Oaks, CA: Sage.

Jaccard, J., & Turrisi, R. (2003). *Interaction effects in multiple regression* (2nd ed.). Thousand Oaks, CA: Sage.

Jerome, G. J., Marquez, D. X., McAuley, E., Canaklisova, S., Snook, E., & Vickers, M. (2002). Self-efficacy effects on feeling states in women. *International Journal of Behavioral Medicine, 9*(2), 139–154.

Kalton, G. (1983). *Introduction to survey sampling.* Thousand Oaks, CA: Sage.

Kehr, H. M. (2003). Goal conflicts, attainment of new goals, and well-being among managers. *Journal of Occupational Health Psychology, 8*(3), 195–208.

Keppel, G. (1991). *Design & analysis: A researcher's handbook* (3rd ed.). Englewood Cliffs, NJ: Prentice-Hall.

Keppel, G., Saufley, W. H., & Tokunaga, H. (1993). *Introduction to design and analysis: A student's handbook* (2nd ed.). New York: W. H. Freeman.

Keyes, C. L. M., Shmotkin, D., & Ryff, C. D. (2002). Optimizing well-being: The empirical encounter of two traditions. *Journal of Personality and Social Psychology, 82*(6), 1007–1022.

Kim, J.-O., & Mueller, C. W. (1978a). *Introduction to factor analysis: What it is and how to do it.* Thousand Oaks, CA: Sage.

Kim, J.-O., & Mueller, C. W. (1978b). *Factor analysis: Statistical methods and practical issues.* Thousand Oaks, CA: Sage.

Kirschenbaum, A., Oigenblick, L., & Goldberg, A. I. (2000). Well being, work environment and work accidents. *Social Science & Medicine, 50,* 631–639.

Klecka, W. R. (1980). *Discriminant analysis.* Thousand Oaks, CA: Sage.

Kline, P. (1993). *The handbook of psychological testing.* London: Routledge.

Knoke, D., & Burke, P. J. (1980). *Log-linear models.* Thousand Oaks, CA: Sage.

Kraemer, H. C., & Thiemann, S. (1987). *How many subjects? Statistical power analysis in research.* Thousand Oaks, CA: Sage.

Lieberson, S. (1985). *Making it count: The improvement of social research and theory.* Berkeley, CA: University of California Press.

Mookherjee, H. N., & Harsha, N. (1997). Marital status, gender, and perception of well-being. *Journal of Social Psychology, 137*(1), 95–105.

Natvig, G. K., Albrektsen, G., & Qvamstrom, U. (2003). Associations between psychosocial factors and happiness among school adolescents. *International Journal of Nursing Practice, 9*(3), 166–175.

Nunnally, J. C., & Bernstein, I. H. (1994). *Psychometric theory.* New York: McGraw-Hill.

Pampel, F. C. (2000). *Logistic regression: A primer.* Thousand Oaks, CA: Sage.

Pedhazur, E. J. (1997). *Multiple regression in behavioral research: Explanation and prediction* (3rd ed.). New York: Holt, Rinehart and Winston.

Philips, M. A., & Murrell, S. A. (1994). Impact of psychological and physical health, stressful events, and social support on subsequent mental health help seeking among older adults. *Journal of Consulting and Clinical Psychology, 62*(2), 270–275.

Rodgers, W. (1995). Analysis of cross-classified data. In L. G. Grimm & P. R. Yarnold, (Eds.), *Reading and understanding multivariate statistics* (pp. 169–215). Washington, DC: American Psychological Association.

Rosenthal, R., & Rosnow, R. L. (1985). *Contrast analysis: Focused comparisons in the analysis of variance.* Cambridge, UK: University of Cambridge Press.

Rosenthal, R., & Rosnow, R. L. (1991). *Essentials of behavioral research: Methods and data analysis* (2nd ed.). New York: McGraw-Hill.

Rosnow, R. L., & Rosenthal, R. (2001). *Beginning behavioral research: A conceptual primer* (4th ed.). Englewood Cliffs, NJ: Prentice-Hall.

Rothman, K. J. (1986). *Modern epidemiology.* Boston: Little, Brown.

Rowntree, D. (2003). *Statistics without tears: A primer for non-mathematicians.* Upper Saddle River, NJ: Pearson Allyn & Bacon.

Runkel, P. J. (1990). *Casting nets and testing specimens: Two grand methods of psychology.* New York: Praeger.

Sayer, A. (1992). *Method in social science: A realist approach* (2nd ed.). London: Routledge.

Schmidt, F. L. (1996). Statistical significance testing and cumulative knowledge in psychology: Implications for training of researchers. *Psychological Methods, 1,* 115–129.

Schumacker, R. E., & Lomax, R. G. (1996). *A beginner's guide to structural equation modeling.* Hillsdale, NJ: Lawrence Erlbaum.

Stevens J. P. (2002). *Applied multivariate statistics for the social sciences* (4th ed.). Hillsdale, NJ: Lawrence Erlbaum.

Tabachnick, B. G., & Fidell, L. S. (2001). *Using multivariate statistics* (4th ed.). Boston, MA: Allyn and Bacon.

Tacq, J. (1997). *Multivariate analysis techniques in social science research: From problem to analysis.* London: Sage.

Valsiner, J., (Ed.), (1986a). *The individual subject and scientific psychology.* New York: Plenum.

Valsiner, J. (1986b). Between groups and individuals: Psychologists' and laypersons' interpretations of correlational findings. In J. Valsiner, (Ed.), *The individual subject and scientific psychology* (pp. 113–151). New York: Plenum.

Ware, J. E. Jr., & Sherbourne, C. D. (1992). The MOS 36-item short-form health survey. I: Conceptual framework and item selection. *Medical Care, 30,* 473–483.

Watson, D., Clark, L. E., & Tellegen, A. (1988). Development and validation of brief measures of positive and negative affect: The PANAS scales. *Journal of Personality and Social Psychology, 54*(6), 1063–1070.

White, J. M. (1999). Effects of relaxing music on cardiac autonomic balance and anxiety after acute myocardial infarction. *American Journal of Critical Care, 8*(4), 220–230.

WHOQOL Group. (1998). The World Health Organization quality of life assessment (WHOQOL): Development and general psychometric properties. *Social Science & Medicine, 46*(12), 1569–1585.

Wickens, T. D. (1989). *Multiway contingency tables analysis for the social sciences.* Hillsdale, NJ: Lawrence Erlbaum.

Wood, W., Rhodes, N., & Whelan, M. (1989). Sex differences in positive well-being: A consideration of emotional style and marital status. *Psychological Bulletin, 106,* 249–264.

Wright, R. E. (1995). Logistic regression. In L. G. Grimm & P. R. Yarnold, (Eds.), *Reading and understanding multivariate statistics* (pp. 217–244). Washington, DC: American Psychological Association.

INDEX

Page numbers in **bold** correspond to key terms within text.

ABOUT THE AUTHOR

John Spicer was associate professor and head of psychology at Massey University, New Zealand, until the end of 2002 when he took early retirement to devote all his time to writing books. Earlier, he was a research fellow for several years at the University of Auckland, New Zealand, and held visiting fellowships at the Universities of Michigan and London. His primary research interests have been in health psychology, and he has published articles mainly on cardiovascular disease and theoretical issues in a variety of international journals. He was coeditor of *Social Dimensions of Health and Disease: New Zealand Perspectives* (1994). Most of his undergraduate and graduate teaching has focused on research methods, particularly multivariate data analysis. In 2002 he coauthored a chapter on sociological and psychological methods in the fourth edition of the *Oxford Textbook of Public Health.*